AN INTRODUCTION TO C++
AND NUMERICAL METHODS

AN INTRODUCTION TO C++ AND NUMERICAL METHODS

JAMES M. ORTEGA
ANDREW S. GRIMSHAW

University of Virginia

New York   Oxford
OXFORD UNIVERSITY PRESS
1999

OXFORD UNIVERSITY PRESS

Oxford   New York
Athens Auckland Bangkok Bogotá Bombay
Buenos Aires Calcutta Cape Town Dar es Salaam
Delhi Florence Hong Kong Istanbul Karachi
Kuala Lumpur Madras Madrid Melbourne
Mexico City Nairobi Paris Singapore
Taipei Tokyo Toronto Warsaw

and associated companies in
Berlin   Ibadan

Published by Oxford University Press, Inc.
198 Madison Avenue, New York, New York 10016

Oxford is a registered trademark of Oxford University Press

Library of Congress Cataloging-in-Publication Data
Ortega, James M., 1932–
    An introduction to C++ and numerical methods / James M. Ortega,
Andrew S. Grimshaw.
        p.   cm.
    Includes bibliographical references and index.
    ISBN 0-19-511767-0 (pbk.)
    1. Numerical analysis—Data processing.   2. C++ (Computer program
language)   I. Grimshaw, Andrew Swift, 1959–   .   II. Title.
QA297.0778   1998
519.4'0285'5133—dc21   97-36115
                                                                   CIP

Printing (last digit): 9 8 7 6 5 4 3 2 1

Printed in the United States of America
on acid-free paper

CONTENTS

PART III Object-Oriented Programming *207*

PREFACE

Since its introduction in the mid 1950s, Fortran has been the primary programming language for scientific computing. However, in the past decade or so, C and C++ have been increasingly used for such work.

This book is not intended to be a textbook for the traditional first course in Computer Science. Rather, it is addressed primarily to students of disciplines such as physics, chemistry, and mechanical, aerospace, and civil engineering, that have traditionally learned Fortran. Many students in these disciplines now have a desire to learn C++, either in addition to Fortran or in place of it. However, they (or their faculty advisors) would still like a course in C++ to concentrate on problems in numerical computation, as opposed to the non-numerical examples typical in the first Computer Science course.

To this end we cover in some detail a number of the basic numerical methods: numerical integration in Chapter 6, solution of a nonlinear equation by Newton's method and the bisection method in Chapter 8, solution of ordinary differential equations by Euler's method in Chapter 10 (but no knowledge of differential equations is assumed), and solution of linear systems of equations by Gaussian elimination in Chapter 13. In addition, a number of other numerical methods are scattered throughout the chapters devoted mainly to programming techniques: approximation of an infinite series, evaluation of polynomials by Horner's rule, matrix-vector multiplication, and so on.

With the exception of the numerical methods, the book follows a more or less conventional organization but there are many variations on how the material can be covered. After the two introductory Chapters 1 and 2, decision statements and iteration are covered in Chapters 3 and 4. However, some might like to postpone Section 3.3 on nested `if` and `switch` statements and Section 4.4 on Advanced Loop Control to a later time. These sections are marked with dagger [†] to indicate they can be skipped at first reading. Functions are covered in two chapters: 5 and 14. The later chapter contains more advanced topics such as a thorough discussion of reference versus value parameters, recursive and inline functions, function overloading, default parameters, templates, and libraries. Any of these topics on functions may be covered earlier, if desired, although arrays should come before reference parameters. Similarly, the data types double, long, char, and various other topics pertaining to arithmetic operations (e.g., type mixing, compound assignment statements) are postponed to Chapter 11 so as not to overload the student too early. But any of these ideas can be covered earlier, if desired.

With the exception of the objects `cin` and `cout`, the book through Part II has been essentially restricted to C for procedural programming. Part III begins an introduction to classes and objects and their use in object-oriented programming, the main difference of C++ from C. A class of complex numbers is used as an ongoing example to illustrate information hiding, constructors, arrays of objects, friend functions, and operator overloading. Then, vector operations and matrix-vector multiplication are used as examples illustrating dynamic memory allocation and the concomitant necessities of destructors and copy and assignment constructors. The book concludes with an introduction to inheritance, derived classes, and abstract base classes. The main example here is an abstract base class for different versions of Gaussian elimination for solving linear systems with either a full or a tridiagonal matrix. For those who wish an earlier introduction to classes, the first chapter (Chapter 17) of Part III can be covered at almost any time, but at least some of the material in Chapter 14 on functions should come first. The remainder of Part III uses dynamic memory allocation, which requires pointers.

We assume this course will be taken mostly by students in the second semester or second year of college. These students are usually taking concurrently the second or third semester of the calculus sequence and thus are familiar with derivatives, integrals, trigonometric functions, etc. and a little linear algebra. Thus the coverage of the numerical methods of the text nicely complements the calculus and linear algebra that the students are currently studying, or will be shortly. The coverage of numerical methods is necessarily elementary at this level, but we believe that it is desirable to stress the importance of errors from the outset. Thus the three major types of errors—rounding error, discretization error, and convergence error in iterative methods—are given fairly detailed discussions, although they cannot be pursued in any depth.

We are indebted to our colleagues Professors Jim Cohoon and Jack Davidson of the University of Virginia for numerous enlightening discussions, and to several reviewers including Professor David Rossomanno of Memphis State University, and students whose comments helped to improve the exposition. Ms. Brenda Lynch has our heartfelt thanks for her expert LATEXing of the manuscript.

AN INTRODUCTION TO C++
AND NUMERICAL METHODS

PART I

BASIC CONSTRUCTS

In this first part of the book, we introduce some of the basic constructs of the C++ language. The goal is to cover enough of the language to allow writing simple, but nontrivial, programs as soon as possible. After a brief introduction to computers and software systems in Chapter 1, we then start in Chapter 2 the discussion of C++, beginning with the basic ideas of variables and their types, assignment, and input/output. Chapters 3 and 4 cover decision and repetition constructs, which are fundamental parts of almost all programs. Then Chapter 5 treats functions, a key mechanism for allowing programs to be broken into smaller parts. Chapter 6 is our first major example in scientific computing: numerical integration. Chapter 7 discusses additional aspects of input and output including the use of files and ways to format output so that it has a pleasing appearance. Chapter 8 considers the solution of nonlinear equations. Chapter 9 introduces arrays, which are extremely important in scientific computing. Finally, Chapter 10 treats our third major scientific computing example: differential equations. There are also discussions of errors and their detection (Section 2.5) and good programming practices (Section 3.4), which can help reduce the number of errors.

In Part II, we will revisit some of the basic C++ constructs of Part I, giving them a more general and detailed treatment, as well as introduce more advanced topics. Parts I and II together cover what is called *procedural programming*, as opposed to *object-oriented* programming, which is the subject of Part III.

1

INTRODUCTION

The many computers now installed in this country and abroad are used for a bewildering variety of tasks: accounting and inventory control, airline and other reservation systems, limited translation of natural languages such as Russian to English, monitoring of process control, and on and on. One of the earliest—and still one of the largest—uses of computers was to solve problems in science and engineering. The techniques used to obtain such solutions are part of the general area called *scientific computing*, and the use of these techniques to elicit insight into scientific or engineering problems is called *computational science* (or *computational engineering*).

There is now hardly an area of science or engineering that does not use computers. Trajectories for earth satellites and for planetary missions are routinely computed. Engineers use computers to simulate the flow of air about an aircraft or other aerospace vehicle as it passes through the atmosphere, and to verify the structural integrity of aircraft. Such studies are of crucial importance to the aerospace industry in the design of safe and economical aircraft and spacecraft. Modeling new designs on a computer can save many millions of dollars compared to building a series of prototypes. Similar considerations apply to the design of automobiles and many other products, including new computers.

Civil engineers study the structural characteristics of large buildings, dams, and highways. Meteorologists predict tomorrow's weather as well as make much longer range predictions, including the possible change of the earth's climate. Astronomers and astrophysicists have modeled the evolution of stars, and much of our basic knowledge about such phenomena as red giants and pulsating stars has come from such calculations coupled with observations. Ecologists and biologists are increasingly using computers in such diverse areas as population dynamics (including the study of natural predator and prey relationships), the flow of blood in the human body, and the dispersion of pollutants in the oceans and atmosphere.

The common denominator in all these diverse areas is that there must always be an algorithm for solving the problem. An *algorithm* is a precise description of the calculations that must be made in order to obtain the solution. An example of a simple algorithm for computing the roots (assumed real) of a quadratic equation $ax^2 + bx + c = 0$ by the formula $(-b \pm \sqrt{b^2 - 4ac})/2a$ is:

1. Compute $u = b^2 - 4ac$
2. Compute $v = \sqrt{u}$
3. Root $1 = (-b + v)/(2a)$
4. Root $2 = (-b - v)/(2a)$

Given the numbers a, b, and c defining the quadratic equation, we could—by pencil and paper or using a calculator—follow the steps of this algorithm to compute the roots. Alternatively, we could instruct a computer to carry out these steps. The main purpose of this book is to set up algorithms that are typically used in solving scientific and engineering problems, and to see how the algorithms may be implemented in the C++ language.

1.1 COMPUTERS AND SOFTWARE

A schematic of a typical computer system is given in Figure 1.1. The main part of the computer consists of the *central processing unit* (CPU) and the *main memory*. These may be connected to a number of *peripheral units*, the most common being a *monitor* and *keyboard*, *disk drives*, a *printer* and possibly other things like a *mouse* and *CD-ROM players*. These peripheral units are called *input/output* (I/O) devices and allow information to be entered into the main computer and retrieved from it. Also, the whole system may be connected to other systems by a *network*.

Suppose that you are presented with such a computer system, perhaps a personal computer (PC). Without certain *software* it would be virtually impossible for you to use this system because the computer will respond only to instructions given to it in the form of binary numbers. A *binary number* is a sequence of the digits 0 and 1, for example, 10110101101, and each computer has an *instruction set* consisting of such binary numbers. These instructions govern the internal operation of the computer as well as use of its peripheral devices. A *machine language program* for a computer

Figure 1.1 A Computer System

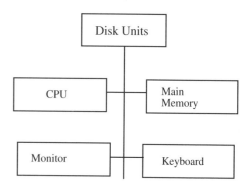

consists of a sequence of binary numbers that instruct the computer to carry out the operations to solve a given problem. In the very early days of electronic computers—in the 1940s—computers were used by writing machine language programs. It was a very tedious and error-prone endeavor.

Operating Systems

One of the main types of *systems software* developed to make the use of computers easier is the *operating system*. An operating system includes commands and facilities for handling input/output, files of data in memory, and various other things. It may be thought of as part of the basic computer system but it is software, not hardware. By means of operating system instructions, certain basic functions may be handled very simply; for example, the instruction

TYPE B:M

may cause the contents of a file named M on disk drive B to appear on the monitor. Without the operating system, a long sequence of binary machine language instructions would be necessary to accomplish this.

Originally, most computer manufacturers developed operating systems for their own products. By and large, the systems were all different—although some general principles pervaded them—and the use of machines from different manufacturers required learning their specific operating systems. Although such vendor-specific operating systems still exist, there has been a strong movement towards general operating systems that can be used on a variety of different machines. The two most important examples of this are the UNIX operating system, developed by Bell Laboratories in the early 1970s, and DOS (Disk Operating System), developed for microcomputers by Microsoft, Inc., in the late 1970s.

Over the past decade, UNIX has increasingly been used in scientific and engineering computation. DOS in conjunction with Windows has been the main operating system for personal computers, but newer versions of Windows are now becoming the standard operating system for PCs.

Programming Languages

Although operating systems provide basic software support, in order to solve actual problems one needs to write programs using a *programming language*. There have been literally hundreds of programming languages developed over the years, but most of them are now only of historical interest. One of the earliest languages was Fortran, developed by IBM in the mid 1950s. Although over 40 years old, it is still a widely used language for scientific and engineering computation. C originally was developed with UNIX in the early 1970s as primarily a "systems" language; in particular, the UNIX operating system is written in C. C++ was proposed about 10 years later as an "object-oriented" extension of C. In recent years, both C and

C++ have been increasingly used for scientific computing. In either C or C++, the algorithm described previously for computing the roots of a quadratic equation can be written as

$$u = b * b - 4 * a * c;$$
$$v = \text{sqrt} (u);$$
$$\text{root1} = (-b + v)/(2 * a);$$
$$\text{root2} = (-b - v)/(2 * a);$$

which is almost self-explanatory (* is multiplication) and mirrors the mathematical algorithm. (We assume in this example that the roots are real so that u is non-negative.)

1.2 BINARY NUMBERS AND MEMORY

The main memory of a computer will hold data, either numbers or characters, and we now begin a discussion as to how this is done. We are used to the decimal number system in which integers are written in terms of powers of 10. Thus,

$$2436 = 2 \times 10^3 + 4 \times 10^2 + 3 \times 10^1 + 6 \times 10^0 \tag{1.1}$$

Here, 10 is the *base* of the number system but any other integer $\alpha > 1$ may also be used as a base. The most common choices for α, besides 10, are 2 (the binary system), 8 (the octal system) and 16 (the hexadecimal system). For any choice of α, integers are written as

$$a_p \times \alpha^p + a_{p-1} \times \alpha^{p-1} + \cdots + a_0 \times \alpha^0, \tag{1.2}$$

where the a_i are integers less than α. The decimal number (1.1) is a special case of (1.2) in which $\alpha = 10$. An example of a binary integer is

$$10110_2 = 1 \times 2^4 + 0 \times 2^3 + 1 \times 2^2 + 1 \times 2^1 + 0 \times 2^0, \tag{1.3}$$

where the subscript 2 denotes that this is a binary number. Examples of octal and hexadecimal integers are

$$6741_8 = 6 \times 8^3 + 7 \times 8^2 + 4 \times 8^1 + 1 \times 8^0, \tag{1.4}$$

and

$$9AF2C_{16} = 9 \times 16^4 + 10 \times 16^3 + 15 \times 16^2 + 2 \times 16^1 + 12, \tag{1.5}$$

where, again, the subscript denotes the base of the number system. In (1.5) we have used the usual hexadecimal convention that

$$A = 10, B = 11, C = 12, D = 13, E = 14, F = 15. \tag{1.6}$$

In the binary system only the digits 0 and 1 are used, in the octal system 0–7 are used, but in the hexadecimal system 0–15 are used, with the latter six represented by (1.6).

The binary number system is used on almost all computer systems today. The hexadecimal system was used on a few systems in the past, and is still very useful as a shorthand for binary numbers, as we now illustrate. The first 16 binary numbers are

$$0000 = 0, 0001 = 1, 0010 = 2, 0011 = 3, 0100 = 4, 0101 = 5, 0110 = 6,$$

$$0111 = 7, 1000 = 8, 1001 = 9, 1010 = 10, 1011 = 11, 1100 = 12, \quad (1.7)$$

$$1101 = 13, 1110 = 14, 1111 = 15$$

and by grouping the digits of a binary number in fours and then using (1.6) and (1.7), we may write it immediately as a hexadecimal number. For example, the binary number 10110110 may be written in hexadecimal as

$$(1011)(0110) = B6 \quad (1.8)$$

Another example is

$$010011101111_2 = (0100)(1110)(1111) = 4EF$$

In the early days of computing, octal numbers were commonly used as a concise way of writing binary numbers, but now hexadecimal is usually used.

Fractions   Just as

$$0.1468 = 1 \times 10^{-1} + 4 \times 10^{-2} + 6 \times 10^{-3} + 8 \times 10^{-4}$$

in the decimal system, fractions can be written in other number systems in a similar way. Thus, for example, a *binary fraction* is of the form

$$0.101101_2 = 1 \times 2^{-1} + 0 \times 2^{-2} + 1 \times 2^{-3} + 1 \times 2^{-4}$$
$$+ 0 \times 2^{-5} + 1 \times 2^{-6}. \quad (1.9)$$

Memory

We next discuss how binary numbers are held in a computer's memory. Memory is usually organized into bytes, where a *byte* consists of 8 binary digits (*bits*). Memory sizes are usually given in terms of bytes. A *kilobyte* (KByte or KB) is 1024 bytes, a *megabyte (MByte or MB)* is 1024 KBytes, and a *gigabyte (GByte or GB)* is 1024 MBytes. Current PCs and larger computers typically have main memory sizes of several Mbytes, and supercomputers have at least a Gbyte. The memory size of hard disk units is usually much larger than that of main memory. Disk memory is slower to access but more permanent than main memory. In addition to main memory and disk memory, most computers today also have *cache* memory, which is much smaller than main memory (typically measured in Kbytes) but much faster to access. In large scientific computing problems it is very important to "manage" these three types of memory properly in order to utilize the computer system efficiently. Figure 1.2 shows schematically the relationship of the various types of memory including *magnetic tape*, which is used primarily for long term storage of data. The cost column pertains to cost per byte.

Integer
Sizes   A byte of memory can hold an integer represented by eight bits so that the largest such binary number is 11111111 or

Figure 1.2 Memory Hierarchy

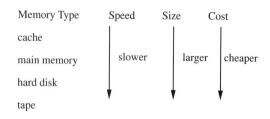

Memory Type	Speed	Size	Cost
cache			
main memory	slower	larger	cheaper
hard disk			
tape			

$$2^7 + 2^6 + 2^5 + 2^4 + 2^3 + 2^2 + 2 + 1 = 2^8 - 1 = 255. \tag{1.10}$$

The first equality in (1.10) is a special case ($\alpha = 2$, $k = 7$) of the identity

$$(\alpha^k + \alpha^{k-1} + \cdots + \alpha + 1)(\alpha - 1) = \alpha^{k+1} - 1,$$

which is verified by multiplying the factors on the left. Integers of the size (1.10) are not large enough for many purposes and usually at least two bytes are used for integer representation. Two bytes allow unsigned (i.e., positive) integers up to $2^{16} - 1 = 65,535$. In some cases, even this is not large enough and some computer systems—either automatically or optionally—use four bytes, allowing integers up to $2^{32} - 1 \doteq 4.3 \times 10^9$. However, for signed integers, one bit is required for the sign so that the maximum magnitude decreases by a factor of 2. For example, with 1 byte only seven bits are available for the magnitude so its maximum size is $2^7 - 1 = 127$. (Actually, negative integers are usually represented by the "two's complement," the details of which need not concern us here; with k bits this allows the magnitude of negative integers to be 2^{k-1} rather than $2^{k-1} - 1$.) Table 1.1 summarizes the maximum sizes of signed and unsigned integers for different numbers of bytes (or bits).

For many computations, integers are not satisfactory, and it is necessary to work with numbers using the "scientific notation" illustrated for decimal numbers by 0.4121×10^6. In binary, such numbers are of the form

$$\pm 0.1 * * \cdots * \times 2^{\pm p}, \tag{1.11}$$

Floating Point Numbers

where each asterisk represents a binary digit, 0 or 1, and $\pm p$ is the exponent of 2. The $.1 * \cdots *$ portion of (1.11) with m bits represents a binary fraction of the form

$$2^{-1} + * \times 2^{-2} + * \times 2^{-3} + \cdots + * \times 2^{-m}. \tag{1.12}$$

An example of (1.12) is given by (1.9). At least 32 bits (4 bytes) are usually used to represent *floating point* numbers of the form (1.11). These 32 bits are divided

TABLE 1.1
Maximum Size of Signed and Unsigned Integers

Number of bytes	Unsigned integers	Signed integers
1 byte = 8 bits	$0 \leq n \leq 2^8 - 1 = 255$	$-2^7 \leq n \leq 2^7 - 1 = 127$
2 bytes = 16 bits	$0 \leq n \leq 2^{16} - 1 = 65,535$	$-2^{15} \leq n \leq 2^{15} - 1 = 32,767$
4 bytes = 32 bits	$0 \leq n \leq 2^{32} - 1 \doteq 4.3 \times 10^9$	$-2^{31} \leq n \leq 2^{31} - 1 \doteq 2.1 \times 10^9$
8 bytes = 64 bits	$0 \leq n \leq 2^{64} - 1 \doteq 2 \times 10^{19}$	$-2^{63} \leq n \leq 2^{63} - 1 \doteq 10^{19}$

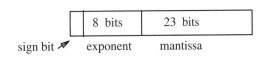

Figure 1.3 Representation of Floating Point Number

between the exponent and the *fractional part* (the asterisks in (1.11)), which is also called the *mantissa*. Figure 1.3 shows a typical representation in memory of a floating point number (this may differ slightly on some computers but the principle remains the same).

The sign bit in Figure 1.3 is 0 if the number is positive and 1 if negative. Eight bits are allotted for a signed exponent, allowing (see Table 1.1) exponents in the range $-128 \leq p \leq 127$; thus, the magnitudes of floating point numbers are in the range 2^{-128} to 2^{127}. The equivalent decimal range is approximately 10^{-38} to 10^{38}. The 23 bits allotted for the mantissa allow binary fractions of the form (1.12) with $m = 24$, since the mantissa is normalized so that its first bit is always 1, and this bit is not actually stored. Since $2^{-24} = 1/16777216 \doteq 10^{-7}$, a 23-bit mantissa can represent approximately seven decimal digits. Thus, using binary floating point numbers with a 23-bit mantissa is roughly equivalent to using decimal numbers with 7 digits. This *single precision* is enough for many computations but sometimes more is needed. *Double-precision* floating point numbers use at least 8 bytes (64 bits) for their representation, with at least 48 bits for the mantissa. This allows the equivalent of at least 14 decimal-digit precision in the numbers. (On some computers, single precision numbers may use up to 64 bits in which case double precision may use up to 128 bits.)

Precision

Characters in Memory

In addition to numbers, the memory may hold characters, which are also represented by binary numbers. *Characters* consist of the letters of the alphabet, both lower and upper case, punctuation marks, and various other symbols. Usually characters are represented in one byte, which allows for 256 different characters (see Table 1.1). There are two common conventions for representing characters: ASCII and EBCDIC. The ASCII (American Standard Code for Information Interchange) convention, which uses only 7 bits, is the more widely used and we will confine our attention to it. For example, the ASCII representation of the letter A in binary is 1000001, which is 65 in decimal. The complete list of ASCII character representations in both binary and decimal is given in Appendix 1.

ASCII Characters

1.3 ROUNDING ERROR

One important source of error in any program that does floating point arithmetic is *rounding error*, which is caused by the fact that computers work with only a finite number of digits. Because of this we cannot, in general, do arithmetic within the real number system as we do in pure mathematics: The arithmetic done by a computer is restricted to finitely many digits, whereas the numerical representation of real numbers can require infinitely many. For example, such fundamental constants as π and e require an infinite number of digits for a numerical representation and can *never* be entered exactly in a computer. Moreover, even if we start with numbers that have an exact numerical representation in the computer, arithmetic operations on them will usually cause errors. In particular, the quotient of two numbers may require infinitely many digits for its numerical representation; for example, $\frac{1}{3} = 0.333\ldots$. And the product of two numbers will usually require twice as many digits as the numbers themselves, for example, $0.81 \times 0.61 = 0.4941$. These numbers will have to be truncated or rounded to the fixed number of digits we are using, for example, 24 bits. Therefore, we resign ourselves to the fact that we cannot do arithmetic exactly on a computer. We shall make small errors, called *rounding* or *roundoff errors*, on almost all floating point operations, and we need to ensure that these errors do not invalidate the computation.

Roundoff errors can affect the final computed result in different ways. First, during a sequence of many operations, each subject to a small error, there is the danger that these small errors will accumulate so as to eliminate much of the accuracy of the final result. It turns out that in most problems this is not as serious as it might *Cancellation Error* seem. More dangerous is the consequence of *cancellation*. Suppose that two numbers a and b are equal to within their last digit. Then the difference $c = a - b$ will have only one significant digit of accuracy *even though no roundoff error will be made in the subtraction*. Future calculations with c may then limit the accuracy of the final result to one correct digit.

Whenever possible, one tries to eliminate the possibility of cancellation in intermediate results by rearranging the operations. We give a simple example of this using three-digit decimal arithmetic. Consider the computation $ab - ac$ for $a = 1.12$, $b = 0.692$, and $c = 0.693$. Then

$$ab = 0.775\underline{04} \quad ac = 0.776\underline{16}$$

where the underscored digits are lost when the results are rounded to 3 digits. Thus the final computation gives

$$0.775 - 0.776 = -0.001$$

as opposed to the correct answer -0.00112. However, suppose we had done the computation as $a(b - c)$. Then

$$b - c = -0.001$$

so that

$$a(b - c) = -0.00112. \tag{1.13}$$

Thus there is no rounding error, and we obtain the correct answer. [This is somewhat misleading, however. Suppose that the "true" values of b and c are $b = 0.69236$ and $c = 0.69294$ so that 0.692 and 0.693 are their best representations with three digits. Then the "true" answer is

$$a(b - c) = 1.12(0.69236 - 0.69294) = -0.00065,$$

which differs considerably from (1.13).]

On the other hand, sometimes cancellation is necessary in order to obtain an accurate answer. Suppose that we wish to compute

$$\sum_{i=1}^{n} a_i - \sum_{i=1}^{n} b_i$$

Cancellation May Be Good

where the a_i and b_i are all positive. If we first sum the a_i and then start subtracting the b_i, we may obtain an inaccurate answer. To illustrate, suppose that we use two-digit decimal arithmetic with no rounding: digits that cannot be retained are dropped. If

$$a_1 = 0.98 \quad a_2 = 0.97 \quad b_1 = 0.95 \quad b_2 = 0.96$$

then

$$a_1 + a_2 = 0.98 + 0.97 = 1.95 \rightarrow 1.9 \text{ (2 digits retained)}$$

Next,

$$(a_1 + a_2) - b_1 = 1.9 - 0.95 \rightarrow 1.9 - 0.9 = 1.0$$

since 0.95 must be truncated to 1 digit to align the decimal point with 1.9. Similarly,

$$(a_1 + a_2 - b_1) - b_2 = 1.0 - 0.96 \rightarrow 1.0 - 0.9 = 0.10$$

However, if we pair the a_i with the b_i, we obtain the exact answer:

$$(a_1 - b_1) + (a_2 - b_2) = (0.98 - 0.95) + (0.97 - 0.96) = 0.04$$

The numerical examples in this section have been done in decimal arithmetic for simplicity, but analogous results hold for binary numbers.

1.4 PROGRAMS

In order to store numbers and characters in memory, and retrieve them later, their locations in memory are specified (much as your street address defines where you live). On many computers each individual byte in memory has an address, but to simplify our discussion we will assume that addresses pertain to each data type. Thus each floating point number, each integer, and each character in memory will have a unique address, given by a binary number. A schematic of the memory is given in Figure 1.4, which also shows typical binary addresses.

Program instructions will also be stored in memory, usually in either two bytes or four bytes for each instruction. (The idea of storing the program itself in memory

Figure 1.4 Data in Memory

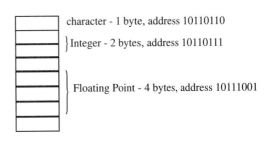

character - 1 byte, address 10110110

} Integer - 2 bytes, address 10110111

Floating Point - 4 bytes, address 10111001

was one of the most important early developments in computers.) A sequence of machine language instructions to add two floating point numbers might be

Opcode	Address

Machine Language Program

0110	10110110 (load number from address 10110110)	(1.14a)
1001	10001101 (add number from address 10001101)	(1.14b)
1101	01101101 (store sum in memory at address 01101101)	(1.14c)

Here, each operation code (Opcode) would consist of a specific binary number for that operation; in this example, there are three different operations, and each would have its own code. Each instruction in this example also contains a binary address of an *operand*, the data used by the instruction. Typically (but not always), instructions will be stored sequentially in memory, but the data will usually be more dispersed. However, the address of a memory location containing data to be used will be in the instructions of (1.14), which allows the correct data to be retrieved.

The CPU

The instructions in (1.14) are carried out by the Central Processing Unit (CPU) shown in Figure 1.1. The CPU consists of two main parts, the *Control Unit* and the *Arithmetic and Logic Unit* (ALU). The Control Unit decodes the current instruction to see what operation is to be done, and initiates a *fetch* (retrieval) of the operand from memory (if required). The ALU contains the arithmetic units that do addition, multiplication, etc. as well as units for logical operations to be discussed later. The CPU also contains a small amount of memory, called *registers*, for temporarily holding instructions and data. In fact, on many computers arithmetic operations obtain their operands only from registers, which must first be loaded from memory.

Assembly Language and Programming Languages

In the very early days of computing, instructions like those of (1.14) would be written out in terms of binary numbers (or octal numbers). Thus the programmer needed to write the binary code for each operation to be performed as well as the binary addresses of all the data in memory. The construction of such *machine language programs* was extremely tedious and error prone. An important early improvement in this process was the development of *assembly language*, in which the instructions in (1.14) could be written in a form like

Assembly Language Program

LOAD	A1	(1.15a)
ADD	B2	(1.15b)
STO	A2	(1.15c)

The *assembler*, the software that converts this program to machine language, would translate the operation symbols ADD, STO, etc., as well as the address symbols, A1, B2, etc., into the correct binary numbers. Although this was a major improvement over writing programs in binary machine language, the programmer had to mirror the machine language code, operation by operation, and also keep track of all the data.

Figure 1.5 Stages of a C++ Program

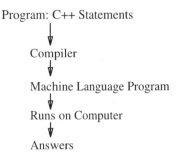

Program: C++ Statements
↓
Compiler
↓
Machine Language Program
↓
Runs on Computer
↓
Answers

The next major advance came with the development of *high-level programming languages*, of which C++ is an example. In C++, the intent of (1.14) [or (1.15)] could be written in the mathematical form

$$a2 = a1 + b2$$

C++ Compiler

The programmer no longer needed to worry about the individual machine operations, nor where data are being stored in memory; this is handled automatically by C++. In order to accomplish this a very important and fairly complex software system is required: the *compiler*. A compiler translates statements written in a programming language like C++ into the machine language code that will actually run on the computer. This is illustrated in Figure 1.5.

The first step of the overall process shown in Figure 1.5 is to write the C++ statements to carry out the desired computation. Much of this book will address this step, and we only note for now that this process will usually involve typing the C++ program at a terminal. The key software used at this stage will be a *text editor*, which is much like a word processor. The editor will allow you to enter statements, correct mistakes, move statements from one position to another, and so on. Some C++ systems have their own editor. In other cases, you may use an editor that is part of the operating system, for example, the *emacs* editor in UNIX. In any case, you will have to learn the details of the particular system you will be using.

MAIN POINTS OF CHAPTER 1

- An algorithm is a set of precise instructions for solving a problem.
- A computer system consists of the central processing unit (CPU), main memory, and various peripherals such as a monitor/keyboard, disk units, and printer.
- Operating systems are basic software that help with input/output, and other functions. UNIX and DOS/Windows are two important general operating systems.
- Numbers are held in computer memory in binary form. Two important types of numbers are integers and floating point numbers. Characters are also represented by a binary code.

- Rounding error contaminates essentially all floating point computation. Especially dangerous are errors caused by cancellation.

- Numbers or characters in memory have unique addresses. These addresses are used in machine language instructions, also stored in memory.

- The CPU consists of the Control Unit, which decodes machine language instructions, and the Arithmetic and Logic Unit (ALU), which does arithmetic and logic operations.

- Assembly language programs mirror machine language but use alpha-numeric symbols rather than binary numbers. They are difficult to write.

- High-level programming languages, such as C++, allow programs to be written in almost mathematical form. Compilers translate such high-level programs to machine language.

- C++ programs are constructed with the help of an editor, which is similar to a word processor.

EXERCISES

1.1 Numbers are represented in computers by

  a. decimal digits

  b. binary digits

  c. characters

1.2 Indicate which of the following are typical of machine language commands

  a. `S = R+T`

  b. `10011001101`

  c. `COPY PR LPT1`

1.3 The main memory of a computer typically holds

  a. data

  b. program

  c. both a. and b.

  d. neither a. nor b.

1.4 Floating point numbers in a computer are binary representations corresponding to which of the following decimal representations:

  a. 0.12×10^4

  b. 1200

1.5 Programming languages such as C++ are useful for:

  a. Doing things one can't do in machine language

  b. Sparing a programmer lots of tedious detail work

1.6 Give a precise algorithm for computing the sum of the squares of n numbers $a_1, \ldots, a_n$.

1.7 Express the following decimal numbers as binary numbers by writing them in terms of powers of 2.

   a. 37

   b. 999

   c. 1024

   d. 0.125

1.8 What is the largest positive integer that can be represented by 5 bits? 7 bits? 9 bits?

1.9 Using three-digit rounded decimal arithmetic, as done in the text, compute $ab - ac$ and $a(b - c)$ for $a = 2.14, b = 0.821, c = 0.712$. Comment on the difference in your answers and how they compare with the exact answer.

2

A FIRST C++ PROGRAM

In this chapter, we first discuss the basic ideas of assignment, variable types, and input/output and use these for our first complete C++ program in Section 2.4. Then in Section 2.5 we consider the important topic of program errors and ways to detect them..

2.1 COMPUTATION AND ASSIGNMENT

Suppose that we want to compute the volume of a sphere of radius r by the formula

$$v = \frac{4}{3}\pi r^3. \tag{2.1}$$

A C++ statement to carry out this computation is

$$v = (4.0 * pi * r * r * r) /3.0; \tag{2.2}$$

which mirrors the mathematical formula (2.1). As is indicated in (2.2), multiplication in C++ is denoted by $*$ and division by $/$; addition and subtraction use the usual symbols $+$ and $-$.

The quantity pi in (2.2) is an approximation to π and must be defined. For example, if we wish π to be accurate to 7 decimal digits, somewhere before the statement (2.2) we could have the statement

$$pi = 3.141593; \tag{2.3}$$

Assignment

The equal sign in (2.2) is the symbol for *assignment*. Given a value of r, and pi defined by (2.3), the computation of (2.2) is carried out using the current value of r, and the computed value is *assigned* to v. r and v are *variables* that correspond to storage locations in memory. Thus the effect of (2.2) is to obtain the current value of r from its storage location, do the computation of (2.2), and put this computed value in the storage location reserved for v.

Since = is used as the symbol for assignment, it is easy to confuse an assignment statement with a mathematical equation. The difference between assignment and mathematical equality is illustrated rather forcefully by the statement

$$v = v+r;$$

Assignment versus Equality

If this were a mathematical equality, it would imply that r equals 0. However, as a C++ statement, its interpretation is that the current values of v and r are added, and this sum replaces the value of v in its storage location. Some other programming languages use different symbols, such as := or ←, to denote assignment, so as not to confuse with mathematical equality.

Another difference between assignment and mathematical equality is illustrated 'by

$$2 = r;$$

rvalues and lvalues

As a mathematical equation, this is perfectly valid, but it is an illegal assignment statement since an assignment cannot be made to the constant 2. Only variables may be on the left-hand side of an assignment statement; such variables are sometimes called *lvalues*. *rvalues* are on the right-hand side of an assignment statement and may be constants, variables, or expressions. Thus in the statement

$$v = 2;$$

v is an lvalue and 2 is an rvalue.

Identifiers

Variables are denoted by *identifiers*, also called *names*, which consist of letters and numbers. For example, instead of r and v we could denote these variables by rad and vol or radius and volume or r1 and v1. As illustrated by these examples, it is good practice to have variable names correspond as much as possible to the mathematical description of the problem. Identifiers can also include the underbar _ ; for example, r_one and v_one might be used. *The first character of an identifier must be a letter, not a number*; for example, 2r would be an illegal identifier. (Although identifiers can also begin with the underbar, this use should be minimized, since such identifiers are used for special purposes.) Also, an identifier must not be a keyword (to be discussed shortly). C++ is *case sensitive*; that is, distinction is made between lower and upper case; for example, VolumeOne and volumeone are different identifiers. (The first form, using capitals for the first letter of multiple-word identifiers, is sometimes preferred over the underbar, as in volume_one.) Although C++ makes no restriction on the length of identifiers, some compilers may limit the length allowable.

Operator Precedence

When we write r * r * r, there is no ambiguity. But with the expression a + b * c we could mean either

$$(a+b)*c \quad \text{or} \quad a+(b*c) \tag{2.4}$$

C++ resolves this ambiguity by the following *operator precedence*:

$$/, \; * \text{ (left to right)} \tag{2.5a}$$

$$+, \; - \text{ (left to right)} \tag{2.5b}$$

That is, the operations /, * are carried out first, left to right, and then the operations + and −, left to right. Thus, C++ would treat a + b * c as equivalent to the second form of (2.4). Some other examples are:

$$a/b * c \text{ is same as } (a/b)*c \tag{2.6a}$$

$$a*b + a/b \text{ is same as } (a*b) + (a/b) \tag{2.6b}$$

In general, parentheses override the precedence so that

$$(a + b) * c$$

gives the other possibility for (2.4).

Parentheses Remove Ambiguity

Even though C++ resolves the above ambiguities by means of the operator precedence (2.5), it is bad practice to rely exclusively on this precedence:

Apparent ambiguities should be resolved by explicitly using parentheses.

This has the advantage that neither you nor someone else reading your program need worry about remembering the precedence, there is less chance for error, and the statements will be clear as written. Thus, for example, you should always use the second forms in (2.6) rather than the first forms.

Ambiguity with Division

With expressions like a * b * c or a + b + c there is no ambiguity and therefore no need to use extra parentheses. However, with a/b/c there are the two possibilities

$$(a/b)/c \quad \text{or} \quad a/(b/c). \tag{2.7}$$

Since C++ evaluates the divisions left to right, the first expression in (2.7) would be computed. Note, however, that

$$(a/b)/c = a/(b*c) \tag{2.8}$$

and since division is many times slower than multiplication on most machines, the second form of (2.8) is preferred. [Some compilers will indeed evaluate a/b/c by means of the second form of (2.8).]

2.2 VARIABLE TYPES

As discussed in Chapter 1, numbers in a computer can take different forms, for example, integer and floating point. Since these forms may require different amounts of memory and also different machine language arithmetic operations, we must declare the type of each variable. We do this by *type definition* or *declaration* statements of the form

Type Declaration

$$\text{int i, j, k;} \tag{2.9a}$$

for integers and

$$\text{float r, v;} \tag{2.9b}$$

for floating point numbers. These declarations must appear before the variables are used.

The variable `pi` of (2.3) should also be defined, and we could include it in the definition (2.9b):

$$\text{float r, v, pi;} \tag{2.10}$$

const

However, a variable such as `pi` should remain constant and not be changed. This may be achieved by the definition

$$\text{const float pi = 3.141593;} \tag{2.11}$$

This definition does three things: It defines `pi` to be floating point; the value of `pi` is set to 3.141593, obviating the need for the assignment statement (2.3); and, finally, `const` says that the value of `pi` may not be changed. An attempt to change it by an

Keywords

assignment would be an error. `const`, `int`, and `float` are examples of *keywords* in C++ and must not be used as identifiers. (We note that it is also possible to define constants by the construction

```
#define pi 3.141593
```

This form, which is common in C, is rarely used in C++ and will not be considered further.)

Just as the value of `pi` was set in (2.11), the value of other variables may be

Initialization

initialized in the type definition statement. For example,

$$\text{int j = 13;}$$
$$\text{float a = 2.12;} \tag{2.12}$$

define `j` and `a` to be integer and floating point variables, respectively, and give them the indicated values. However, since `const` is not used with these variables, they may be changed by later statements. Initialization may also use the alternative form

```
int j(13);
float a(2.12);
```

but we will use only the style of (2.12) until we discuss classes in Part III.

It is always a good idea to define important constants of the problem by a statement such as (2.11). While it would be permissible to not define a constant `pi`

Define Constants of the Problem

at all and simply write (2.2) as

```
v = (4.0 * 3.141593 * r * r * r)/3.0;
```

this would tend to obscure the role of `pi` as a key constant. And in other situations, a constant may not be so easily identified. For example, the purpose of the statement

```
float price = 13 * 0.30;
```

is not nearly as clear as with the statements

```
const float number_of_oranges = 13;

const float price_per_orange = 0.30;

float total_price = number_of_oranges * price_per_orange;
```

This last statement also illustrates that variables may be initialized by expressions containing other variables, provided that these other variables have been assigned values.

Beware of Integer Division

All the arithmetic operations `*`, `/`, `+`, `-` apply to integer variables as well as floating point. However, if `i` and `j` have been declared to be `int`, the division `i/j` gives only the integer part of the quotient. For example,

$$17/6 = 2, \ 1/2 = 0, \ 45/10 = 4$$

Thus be very careful of integer division and use it only when the integer part of the quotient is what you actually wish.

There is also an operation, `%`, that gives the remainder of an integer division; for example,

$$17 \ \% \ 6 = 5, \quad 1 \ \% \ 2 = 1, \quad 45 \ \% \ 10 = 5$$

2.3 INPUT AND OUTPUT

At this point of our discussion, a C++ program to compute the volume of a sphere would include the statements (2.9b), (2.11), and (2.2):

```
float r, v;

const float pi = 3.141593;

v = (4.0 * pi * r * r * r)/3.0;
```

However, we have not yet provided a value of `r` for which the computation is to be done. We could use another statement to assign a value to `r`, for example,

```
r = 2.4;
```

cin and cout

But this is very restrictive; we would like the ability to have different values of `r`, rather than setting a fixed value. For this purpose we will use the *input statement*

$$\text{cin} \ >> \ \text{r;} \qquad\qquad (2.13)$$

which allows a value of `r` to be entered at the keyboard. Then, after the computation of `v` is complete, if we wish to see what has been computed, the *output statement*

$$\text{cout} \ << \ \text{v;} \qquad\qquad (2.14)$$

will display the value of v on the monitor. Thus by means of (2.13) and (2.14) we can enter data at the keyboard and show the computed results on the monitor. We may also use expressions with cout, as in

$$\text{cout} \ll (\text{v} + 2.0);$$

Here, parentheses should enclose the expression, as indicated. cin and cout are pronounced "see in" and "see out," respectively.

Streams   In the language of C++, a *stream* is a flow of characters or other data. An *output stream* is a flow out of the program. << is called the *insertion operator*, and a cout statement *inserts* values into the output stream. Likewise, an *input stream* is a flow into the program. >> is the *extraction* operator, and a cin statement *extracts* values from the input stream. This is depicted in Figure 2.1.

Prompt   When cin >> r is executed, the computer will halt until a value for r is entered from the keyboard. In order to indicate that this halt has occurred, it is good practice to precede this cin statement by a *prompt* statement of the form

$$\text{cout} \ll \text{"r = ?";} \qquad (2.15)$$

In (2.15), the quotation marks signify that what is enclosed between them is to be printed as is, including any blank spaces. Thus, when (2.15) is executed, r = ? will appear on the monitor screen. If this statement immediately precedes the cin >> r statement, the appearance of this message on the screen indicates that now is the time to enter a value of r from the keyboard.

Labels   Constructions similar to (2.15) are also useful for labeling the results of a computation. For example, the statement

$$\text{cout} \ll \text{"v = " } \ll \text{v};$$

will produce a line on the monitor that looks something like

$$\text{v} = 4.18879$$

Cascading   This cout statement has two insertion operators and a similar construction may be used with cin; this is called *cascading*. For example, the statement to read values of two variables r1 and r2 could be

$$\text{cin} \gg \text{r1} \gg \text{r2};$$

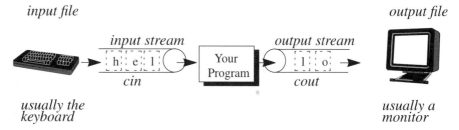

Figure 2.1 Input and Output Streams

Here, either a space or carriage return is given after typing the first number on the keyboard, before typing the second one.

2.4 A COMPLETE PROGRAM

Comments

Let us now put together the various statements we have discussed so far in order to have a complete program, as shown in Example 2.1. The first statement of this program is a *comment statement*. Although this statement is optional, it is good practice always to start a program with a comment statement that says what the program does. A comment statement, as signified by the //, is used to give information to a human reader and has no effect on the program. As illustrated in the program, these comments may be separate lines or may be on the same line as executable statements. In general, anything on a line following // is interpreted as a comment and is ignored by the compiler. Note that there can be no space between the two slashes. It is good practice to use comments throughout the program to indicate what is being done.

EXAMPLE 2.1
A C++ Program for Computing the Volume of a Sphere

```
// This program computes the volume of a sphere.
// It receives the radius from the keyboard, computes
// the volume and displays it on the monitor.
// Author: J. Student
// Date: 7/1/97
#include <iostream.h>
void main(){
  float r,v; // declare variables.  r is radius.  v is volume
  const float pi = 3.141593;     // set value of pi
  cout << "r = ?\n";             // prompt for read of r
  cin >> r;                      // read r from keyboard
  cout << "r =" << r << "\n";    // echo the input
  v = (4.0 * pi * r * r * r)/3.0;  // do the computation
  cout << "v = " << v << "\n";   // display the volume v
} // the program now ends
```

Directive

The statement following the first set of comment statements in Example 2.1 is a *preprocessor directive* that says this program is to use a program from a file called iostream.h in conjunction with the input/output commands cin and cout. This statement *must* be present whenever cin or cout are used. Note that there can be no space between < and iostream.h. (As will be discussed in detail in Part III, cin and cout are objects of the class iostream.)

Program Body

The next statement, void main(), denotes the beginning of the program proper; we will discuss the meaning of void later. The program following main() is enclosed in { }, and one convention is to locate these braces as shown. [Some writers prefer to have the first { left-justified on a separate line following main().] The statements between these braces are called the *body* of the program. The first statement in the program body defines the type of the variables in this program to be floating point. Next follows the definition of pi, and the prompt and read statements.

New Line

The \n in the cout statement says to move to a new line on the monitor after r = ? is printed. The \ in this context is called an *escape character*. It indicates that the following n is not the letter n but is the *new line character*. Another use of the escape character \ is when you wish to display quotes. For example, if you wished to display "T", with the quotes, the statement

```
cout << ""T"";
```

Escape Character

will not work. Instead you need to write

```
cout << "\"T\"";
```

Here, the escape character \ indicates that the following " is to be treated as another character and not as the closing " to terminate the string of characters.

Echo

We have added after the cin statement in Example 2.1 a display of the input r. This is called an *echo* of the input, and it is good practice to include this display to ensure that your input is correct. The computation of v follows, and then the output of v to the monitor screen.

The program in Example 2.1 illustrates a number of rules and conventions in writing C++ programs:

- Every executable statement or declaration is terminated by a semicolon. Note that, however, main(), the directive #include <iostream.h> and the ending brace of the program are not followed by semicolons.

- \n must be enclosed in quotes, perhaps as part of a larger string of characters, as in the first cout statement.

- Each new entity in a cout statement is preceded by <<. Thus, in the last cout statement there are the three entities "v = ", v, and "\n", each preceded by <<.

- At least one space must separate key words from variables, as in float r. Additional spaces are ignored, as are blank lines and tabs. It is useful for better readability to have one space on each side of operators (*, <<, and so on) and =.

- All keywords (float, const, include, etc.) and other words such as cout and cin must be in lower case.

endl

We note that endl (the last character is the letter l) may be used rather than \n for line spacing. For example, the first cout statement in Example 2.1 may be written as

```
cout << "r = ?" << endl;
```

flush

As opposed to \n, endl is not enclosed in quotes and is preceded by <<. (endl is slightly less efficient than \n, but this will make no difference on most programs.) A related command is illustrated by

```
cout << "r = ?" << flush;
```

Here, `flush` says to "flush the buffer" that is holding `r = ?`. This displays `r = ?` but doesn't go to a new line. This is sometimes used when you wish the input value for `r` to go on the same line as the prompt. However, it is redundant in this situation since the statements

```
cout << "r = ?";
cin >> r;
```

will also allow typing `r` on the same line that the prompt `"r = ?"` is printed. The reason is that `cout` is *tied* to `cin` and the buffer is automatically flushed before the `cin` operation begins. We note that `endl` flushes the buffer and positions for a new line, whereas `\n` positions for a new line but does not necessarily flush the buffer.

Entering and Running the Program

- Once you are in your editor environment, type in the program and store it in a file named, say, `volume.cpp`.
- Now execute whatever commands your system may require to compile and execute the program.

Various errors may occur in either the compilation or in the execution of the program; these will be discussed shortly. Assuming the program has compiled and run correctly, you will see on your monitor something like

$$r = \ ? \quad \text{(prompt)}$$
$$1.0 \quad \text{(entered value)}$$
$$r = 1 \quad \text{(echo of input)}$$
$$v = 4.18879 \quad \text{(output value)}$$

You have now run your first C++ program!

We note that some systems may perform the output using "scientific notation" such as $0.418879e + 12$ for 0.418879×10^{12}.

2.5 ERRORS AND DEBUGGING

It is quite common to make errors in writing programs. These are generally of four types: syntax errors, link errors, run-time errors, and logic errors. A program that has errors is said to have *bugs*, and the process of detecting and correcting these errors is called *debugging*.

Syntax errors occur when the rules of the language are violated. For example, trying to use `2r` as a variable name is a syntax error, since, as mentioned previously, `2r` is an illegal identifier. Other examples are: Whenever there is an opening `"`, as

Syntax and Link Errors

in (2.15), there must be a closing `"`; and if instead of `r * r`, you mistakenly typed `r ( r`, this would be an unknown construction to the compiler. Syntax errors are usually the easiest to correct since compilers will detect them and give (sometimes rather cryptic) error messages. You may need to see your local compiler manual for further details on the error messages your compiler produces. *Link errors* may occur at the end of the compilation process because the *linker* (to be discussed in Chapter 14) cannot find some information that your program requires.

Run-time errors (which include *data-dependent* errors) occur after the program has compiled correctly and execution is attempted. One frequent source of such errors is that the data are not suitable. For example, we noted in Chapter 1 that if 32 bits are allowed for floating point numbers, the maximum such number is about 10^{38}. If we entered the value of r as 10^{15} in the program of Example 2.1, then r^3 would exceed this allowed magnitude; this is called an *overflow* and is a run-time error. (If we had entered $r = 10^{-15}$, r^3 would be less than the smallest allowed number of approximately 10^{-38}; this is called an *underflow*. In this case, most systems will just set the result to zero and not give a run-time error.) Other common types of run-time errors are attempts to take the square root of a negative number or to divide by zero. When a run-time error has occurred, most systems will give an error message, similar to compiler-error messages, that will usually make it clear what the problem is. Probably the most common run-time error message you will see is "general protection fault" (called "segmentation fault" on UNIX systems), which can be caused by things we will discuss later. Note, however, that some systems may not give any indication of a run-time error and the final results will simply be erroneous. For example, on some systems overflow of integers will not give a run-time error indication; if a computed integer result is larger than the number of bits allowed for integers, the result will simply be erroneous.

Logic errors are errors in formulation: The program will compile and run without error messages, but it does not accomplish what you intended. For example, suppose that instead of the correct formula for the volume of a sphere you mistakenly typed

```
v = (4.0 * pi * r * r)/3.0;
```

neglecting the final `* r`. The program will now compute the volume incorrectly. The compiler cannot help in detecting this type of error since it doesn't know what you are trying to do.

Another type of error is the following. Suppose you wished to read from the keyboard values for two variables `u` and `v` and wrote

```
cin >> u, v; //incorrect
```

This statement will give neither a syntax error nor a run-time error indication, but only a value for `u` will be accepted from the input stream. As discussed earlier, the correct statement is

```
cin >> u >> v;
```

Logic errors are usually the most difficult type of error to detect. A first check is

Run-Time Errors

Overflow
Underflow

Other Run-Time Errors

Logic Errors

always *consistency*: Are the computed results reasonable for the problem at hand? For example, if you are computing the volume of a sphere and it is negative, there is an error. Or, if you are computing rocket trajectories that you expect to attain a height of 5 kilometers and your program is showing 500 kilometers, obviously something is wrong.

Test Cases

Even if your program is computing reasonable results, that doesn't mean they are correct. You should always run your program on several *test cases* for which you know the answer. For example, in the context of the sphere program of Example 2.1, you should calculate the answer on a hand calculator for a few values of r and verify that your program reproduces these answers. The more test cases you run successfully, the more your confidence in the correctness of the program.

An important point is that many programs fail for very special conditions that may not even be of much interest. For example, in a later chapter we will study the solution of n linear equations in n variables. It is possible to write programs for this problem that will fail for the trivial case $n = 1$. Thus programs should also be tested for as many special-case situations as possible.

Temporary Prints

Almost all compilers will have an associated *debugger*, which will help you to find errors in your program. Amongst other things, debuggers will allow you to stop your program at intermediate points and let you examine current values of variables. For example, suppose you suspect that there is an error in some particular part of the program and you wish to examine what the results are at the end of some intermediate statement. To illustrate, we give in Example 2.2 a code fragment to display the current values of three variables k, i, and m. The cout statement as well as the two following statements are not part of the program and have been added temporarily for debugging purposes. The cin statement in Example 2.2 causes the program to halt so you can examine the results on the monitor screen; entering a value for the temporary variable j and hitting the enter key will resume operation. If you are unable to use a debugger, you may add corresponding temporary statements in your own program. Once the program is correct, you will remove the cout and cin statements.

EXAMPLE 2.2
A Temporary Print

```
    k = k + i * m;
//This is a temporary print and pause
    cout << i << k << m << endl;
    int j;
    cin >> j;
```

Commenting Out

It is sometimes useful to make comments out of certain statements in the program so as to deactivate them temporarily. For example, one way to try to locate a run-time or logic error is by deleting parts of the program until you find the problem. Commenting out parts of the program is preferable to deleting, since you will not have to retype the deleted portions once the error is found.

Suppose, for example, in the sphere program of Example 2.1, we are getting erroneous results and we wish to see what the output is in a trivial case. We might then make a comment of the computation statement:

```
// v = (4.0 * pi * r * r * r)/ 3.0;
```

and insert the statement

```
v = 1.0;
```

to see if 1.0 is displayed for v. Later, when we are ready to do the real computation, we would remove the // from the computation statement and also remove the statement `v = 1.0;` .

It is doubtful that one would ever comment out statements in a program as simple as that of Example 2.1, but in larger and more complex programs it can be a useful technique. You may comment out several consecutive lines of a program by putting `/*` at the beginning of the first line and `*/` at the end of the last line. This may be preferable to putting // before each statement.

MAIN POINTS OF CHAPTER 2

- Arithmetic expressions use the symbols /, *, +, and – for division, multiplication, addition, and subtraction.
- Division of an integer by an integer yields only the integer part of the quotient.
- The arithmetic operators * and \ have higher precedence than + and – and are carried out first. Within each precedence class, operations are performed left to right.
- Statements such as z = x + y assign values of a computation to a variable. Statements of this kind mirror mathematical formulas.
- Identifiers consist of letters or numbers, the first one a letter. An underbar may also be used.
- All variables must be explicitly declared. Initialization may be included in declarations.
- A const declaration may be used for a variable that remains constant.
- The cin statement allows input from the keyboard, and the cout statement displays results on the monitor.
- cin statements should be preceded by cout statements as prompts and followed by cout statements that echo the input.

- Programs that use `cin` and `cout` must contain the directive

```
#include <iostream.h>
```

- Comments should be used to describe the program.

- A program may contain syntax, link, run-time, and logic errors. A program should be extensively tested to detect any errors. Some tools for error detection are debuggers, and when a good debugger is not available, temporary print statements and commenting out sections of code.

EXERCISES

2.1 Write C++ statements for the following computations:

  a. $a = r^2$

  b. $s = a^2 + b^2 + c^2$

  c. $w = [(x - y)^2 - (x + y)]/32.0$

  Give declaration statements for all of the variables, assuming that those in part b are integer and the others are floating point.

2.2 Indicate which of the following are legal identifiers. Indicate which are illegal, and why.

  a. `feb29`

  b. `29feb`

  c. `feb.mar`

  d. `_feb_mar`

2.3 If i and j are `int` variables with the values i = 2 and j = 3, what are the values of the following expressions?

  a. `i/j`

  b. `i%j`

  c. `(j - i)/(j + i)`

2.4 Insert parentheses into the following expressions such that the resulting order of operations is the same as determined by the precedence (2.5).

  a. `a + b/c + d/a * c`

  b. `a * b/c/d * a + c`

2.5 What is the effect of the following statements?

```
const float r = 4.0;
r = r * r;
```

2.6 Write a complete program to display the message "Hello World" on the monitor. (This is a standard first example program in many C++ books.)

2.7 Run the program of Example 2.1 for various values of r. Verify that your results are correct.

2.8 Modify the program of Example 2.1 so as to compute the area of a circle by the formula $A = \pi r^2$.

2.9 Write complete programs to carry out the computations of parts b. and c. of Exercise 2.1. Declare all variables, input the variables by cin statements preceded by prompts, echo the input, use cout statements with suitable labels for the output, and use comments to describe the program.

2.10 Consider the program

```
#include <iostream.h>
void main(){
   int k, m, n;
   cin >> n >> m;
   k = (n * n) + m * m * m;
   cout << k;
}
```

If values of 4 and 3 are read for n and m, respectively, what value is printed for k?

chapter

3

DECISIONS, DECISIONS

One of the most important facets of programming is the ability of a program to take different courses of action depending on input to the program or on current calculated values. Which course of action to take is decided by a *conditional statement*, which in C++ is the `if` statement. In this chapter, we introduce the `if` statement in its various forms, as well as other constructs for making decisions.

3.1 THE `if` STATEMENT

Suppose that a program contains the statement

$$q = b/a; \qquad\qquad (3.1)$$

In order to ensure that you never divide by zero, you can precede the statement (3.1) with a test for zero, as shown in Example 3.1. In general, it is good practice to have a program check its data for any condition that could cause an error before proceeding with a computation.

EXAMPLE 3.1
Program Fragment for `if` Statement

```
if(a == 0)
    cout << "error: a is 0";
else
    q = b/a;
```

The program fragment of Example 3.1 implements the logical diagram shown in Figure 3.1. In particular, if a is zero, the following statement is executed so that an error message is printed and the statement following `else` is bypassed. If a is not zero, the `cout` statement is bypassed, and the statement following `else` is executed.

Figure 3.1 Flow Chart
Corresponding to
Example 3.1

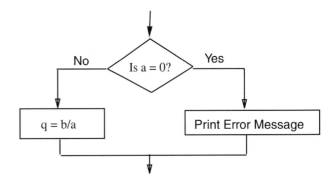

Figure 3.1 is a very simple example of a *flow chart* for outlining the logical flow of a program. Flow charts will be used throughout the book. As shown in Example 3.1, the cout and q = b/a statements are indented; this is not necessary but is considered good practice in order to make a program easier to read.

Flow Chart

The *relational* or *comparison operator* == used in the if statement is one of six such operators in C++, as summarized in Table 3.1.

Relational Operators

Example 3.2 gives a complete program that illustrates the if statement in another context. This program computes the value of the discontinuous function

$$f(x) = \begin{cases} x^2, & \text{if } x < 0 \\ x + 2, & \text{if } x \geq 0 \end{cases} \tag{3.2}$$

EXAMPLE 3.2
The if Statement for a Discontinuous Function

```
#include <iostream.h>
void main(){
  float f, x;
  cout << "what is x?\ n";
  cin >> x;
  if(x < 0)
    f = x * x;
  else
    f = x + 2.0;
  cout << "The value of f is " << f << endl;
}
```

TABLE 3.1
Relational Operators in C++

Mathematics	$=$	$\leq$	$\geq$	$<$	$>$	$\neq$
C++	==	<=	>=	<	>	!=

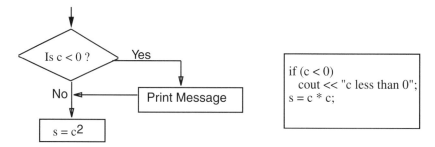

Figure 3.2 Another Flowchart and Program Fragment

The flow chart of Figure 3.1 indicates a situation in which we wish to do either one thing or another, and then continue the program. Another possibility is depicted in Figure 3.2, in which an additional statement is executed in case the tested condition holds. The logic of the flow chart of Figure 3.2 can be implemented by an if statement that does not include the else part, as also shown in Figure 3.2.

The code fragment in Figure 3.2 may also be written in the condensed form

```
if(c < 0) cout << "c less than 0";
s = c * c;
```

Blocks

This form may be used when only a single statement will be executed if the test is true. If more than one statement is to be executed in either the if or the else clause, these statements must be enclosed in braces to form a *block*, as illustrated in Example 3.3. This example illustrates one conventional placement of these braces in an if statement. In general, arbitrarily many statements may be in braces following the if statement, and also following the else. Example 3.3 also illustrates that an arithmetic expression may be used in the test in the if statement. Thus, for example, if the value of a is 0.9 when the if statement is executed, a * a = .81 < 1 so that the following two statements are executed.

EXAMPLE 3.3
An if Statement with a Block

```
if ((a * a) < 1.0){
    a = a + 1.0;
    b = b + 1.0;
}
```

The exit **Statement**

If an error condition occurs in a program, such as a == 0 in Example 3.1, there may be no reason to continue the program. In this case, we could modify Example 3.1 as shown in Example 3.4.

EXAMPLE 3.4
Terminating Program If an Error Occurs

```
if(a == 0){
  cout << "error: a = 0";
  exit(1); //requires stdlib.h
}
else
  q = b/a;
```

The statement `exit(1);` in Example 3.4 causes the program to terminate and return control to the operating system. The number 1 indicates to the operating system that an error has occurred in the program. To make this clear, the statement can also be written as

```
exit(EXIT_FAILURE);
```

On occasion, you may wish to exit a program even though there has been no error. In this case, the `exit` statement may be written as

```
exit(EXIT_SUCCESS); or exit(0);
```

In order to use the `exit` statement, the directive

```
#include <stdlib.h>
```

must be included at the beginning of the program.

Testing on 0 or 1

= versus == in Test   A common error is to write = in a test rather than the correct ==. C++ does not consider = to be a syntax error, and this can lead to bizarre behavior of the program. When the statement

```
if(a == 3) cout << "Help";
```

is executed, if a is 3, a nonzero value (perhaps 1) is assigned to the expression `a == 3`, whereas if a is not 3, a value of zero is assigned. The `if` statement then tests these expression values for zero or nonzero to make its decision. If the above statement were mistakenly written as

```
if(a = 3) cout << "Help";
```

the value 3 is assigned to a, and also to the expression to be tested by the `if` statement; consequently, the expression is nonzero and will always be true. But if the expression were a = 0, the value 0 would be assigned, and the expression would always be false.

3.2 LOGICAL OPERATORS

In the previous section we introduced the `if` statement in the context of a simple decision of the form

```
if(a == 0)   cout  <<   a;
```

More generally, we may have

```
if(logical  expression) cout  <<   a;
```

where *logical expression* (also called a *conditional expression* or *boolean expression*) will take on values of either true or false, represented by nonzero or zero. (Strictly speaking, C++ does not distinguish logical expressions from other expressions: They are just expressions. However, we prefer to make the distinction that these expressions may be thought of as expressing true or false.)

Logical Operators and (&&) or (||) not (!)

The most common way of constructing logical expressions is by means of the relational operators ==, <, and so on, shown in Table 3.1, and the *logical operators* and, or, and not, which are represented in C++ by &&, ||, and !, respectively. For example, consider the statement

```
if((a == 0)  ||  (b < 2.0))   cout << a << b;                 (3.3)
```

If either `a == 0` or `b < 2.0` when this statement is encountered, then the `cout` statement will be executed. Each of `a == 0` and `b < 2.0` is an expression taking on values true or false, and their combination with the || operator gives a more complicated expression. We note that the inner parentheses in (3.3) may be removed since the comparison operators, == and <, take precedence over the operator ||. However, these parentheses help readability.

Two more examples are

```
if(a == 0  &&  b < 2.0)   cout << a;                          (3.4)

if(!(a == 0))   cout << a;                                    (3.5)
```

In the first case, the logical expression takes on the value true if and only if both `a == 0` and `b < 2.0` are true. The second case, (3.5), is equivalent to the statement

```
if(a  ! = 0) cout << a;                                       (3.6)
```

The statement (3.6) would usually be preferable to (3.5), although the ! operator is sometimes very natural. The values of the logical operators are summarized in Table 3.2, where p and q denote logical expressions. (Another way to express the content of Table 3.2 is given in Exercise 3.6.)

If we have an expression with two or more logical operators, for example,

```
a > 0  || a < 10  &&  b == 1                                  (3.7)
```

there is a potential ambiguity: Does this expression mean

```
(a > 0  || a < 10)  &&  b == 1                                (3.8)
```

TABLE 3.2
Values of Logical Operators

Expression	Value
!p	True if p is false and false if p is true
p && q	True if both p and q are true
p || q	True if either p or q (or both) is true

or

$$a > 0 \;\;||\;\; (a < 10 \;\;\&\&\;\; b == 1) \tag{3.9}$$

They are not the same since if $b \neq 1$ and $a = 20$, (3.8) is false, but (3.9) is true. C++

Precedence resolves this ambiguity by the precedence (left to right):

$$\text{First} \quad ! \quad \text{Then} \;\&\&\; \text{Then} \;||\; \tag{3.10}$$

Thus, in (3.7), && is performed first so that (3.7) is equivalent to (3.9). As in arithmetic expressions, parentheses override the precedence (3.10) so that (3.8) is an expression different than (3.7). The comparison operators themselves have a lower precedence than arithmetic operators so that with $a > b + c$ the expression $b + c$ is evaluated first. However, *use of parentheses is recommended in situations like this to improve readability.*

Short Circuit
Evaluation
In (3.4) the expression $a == 0$ is evaluated first, and if it is not true, then the whole expression must be false and the evaluation of $b < 2$ may be bypassed; this is called "short circuit" evaluation. In general, the leftmost expression is evaluated first, and the second expression need not be evaluated if the first expression is true and the operator is || or if the first expression is false and the operator is &&.

3.3 NESTED if AND Switch STATEMENTS†

Consider the flow chart in Figure 3.3. This depicts a logical flow that cannot be implemented with a single if statement. It can, however, be implemented by an if

Figure 3.3 A Logical Flow Chart

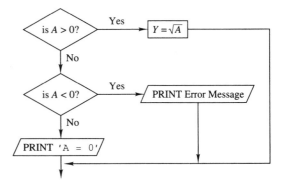

---

†This section may be omitted on first reading.

statement within an `if` statement, called a *nested* `if` *statement*. This is illustrated in Example 3.4(a).

EXAMPLE 3.4
Nested `if` for Figure 3.3

```
if(A > 0)                          if(A > 0)
  y = sqrt(A);                       y = sqrt(A);
else                               else if(A < 0)
  if(A < 0)                          cout << "help";
    cout << "help";                else
  else                               cout << "A = 0";
    cout << "A = 0";
  (a) Nested if                       (b) Else if form
```

The nesting of `if` statements may be repeated to have an `if` within an `if` within an `if`, and so on. However, deep nesting of `if`s may be difficult to comprehend when they are written in the fashion of Example 3.4(a). An alternative way to write the nested `if` statement of Example 3.4(a) is shown in Example 3.4(b). This is exactly the same nested `if` statement as in Example 3.4(a) but with `if(A < 0)` moved to the `else` line and the indentation changed.

else if Form

The "`else if`" form of a nested `if` statement is especially useful when there is deep nesting. The general `else if` form is shown in Example 3.5, in which arbitrarily many `else if` clauses are indicated. The statements following the first logical expression that is true are executed, and control is then passed to the statement following the whole `if` statement. If none of the logical expressions is true, then the statements following the optional `else` statement are executed. Thus the `else if` form provides a way to make a nested `if` statement more easily understandable. Another example of the `else if` form is given in Exercise 3.8. We note that if only a single statement follows an `if`, `else if`, or `else` in Example 3.5, then braces surrounding the statement are not necessary, as in Example 3.4.

EXAMPLE 3.5
The General `else if` Form of Nested `if`s

```
if(logical expression)
      {statements}
else if(logical expression)
         ⋮
else if(logical expression)
      {statements}
else
      {statements}
```

The Switch Statement

EXAMPLE 3.6
A `switch` Statement
_____

```
switch(n){
  case 1: a = 1;
    break;
  case 3: b = 2;
    break;
  case 6: c = 4;
    break;
}
```
_____

The selection of one of several possible courses of action by a nested `if` or `else if` sequence may also be achieved more efficiently by the `switch` statement, in situations where it is applicable. Consider the code shown in Example 3.6, which is equivalent to

```
if(n == 1)
  a = 1;
else if(n == 3)
  b = 2;
else if(n == 6)
  c = 4;
```

Can't Use Switch with `float`

Thus, if the *selector variable* n is 1, 3, or 6, the statement in the corresponding case block is executed. Then the `break` statement passes control to the statement following the switch statement so that the other cases are bypassed. If n is not one of these integers, all cases are bypassed; therefore, at most one `case` block is executed. In Example 3.6, it assumed that n is an integer variable. The `switch` construct may not be used with a floating point selector variable.

The general form of the `switch` statement is shown in Example 3.7. In each case in Example 3.7, a block of statements may be given, not just a single statement. Even if there is more than a single statement, braces are not necessary to enclose the block. Also, there may be an optional `default` case, which is executed if no other case is. We note that each `value` must be a constant, as were 1, 3, and 6 in Example 3.6, and not a variable.

In the previous examples, the `break` statements ensure that only a single case is executed. If the `break` is eliminated, the `switch` will proceed to the next statement so that more than one case block will be executed.

The Assert Statement

We end this section with two other constructs that use logical expressions and are occasionally useful for debugging, as well as other purposes. The first is the

EXAMPLE 3.7
The General Switch Construct

```
switch(selector variable){
    case value:
        statements
        break;
    case value:
        statements
        break;

            ⋮

    case value:
        statements
        break;
    default:
        statements
}
```

`assert` statement. Consider, for example, the statement

$$assert(r > 0);$$

If `r` is not positive when this statement is executed, then an error message will be issued and the program will halt. More generally, any logical expression may be used in place of `r > 0`. Thus, whenever you wish to ensure that some condition is satisfied, you may put in an `assert` statement at the appropriate point. In order to use `assert` statements, you must add the directive

$$\texttt{\#include <assert.h>}$$

at the beginning of the program.

If you are using `assert` statements only for debugging, when the program is correct, you may wish to delete them. Alternatively, you may leave them in the program but deactivate them by adding the statement

$$\texttt{\#define NDEBUG}$$

before the `#include <assert.h>` statement. Then, if at a later date you discover more errors in the program, you may reactivate the `assert` statements simply by removing the NDEBUG directive.

Conditional Compilation

A sometimes useful adjunct to temporary prints (Section 2.5) is the idea of *conditional compilation*, as illustrated by the following program skeleton:

```
#include <iostream.h>
const int test = 1; //set test parameter
```

```
void main(){
  //code here
#if test
  //code here that compiles only if test != 0
#endif //end of the conditional code segment
  //more code here
```

The code included between `#if test` and `#endif` compiles and executes only if the parameter `test` is nonzero; if `test` is zero, this segment of code is not compiled and is not part of the program. This conditional code might contain `cout` statements that display values of some intermediate variables when debugging the program. Then, when the program is running correctly, this section of the code may be made inactive by redefining the parameter `test` to be zero. The advantage of this approach over inserting temporary print statements that will be removed once the program is debugged is that the testing code is available if further changes to the program are made that require additional debugging.

3.4 GOOD PROGRAMMING PRACTICE

One of the biggest mistakes in programming is to begin writing the code before doing adequate planning. We should first make sure that we understand the "big picture" before attempting to write the final C++ statements. This is especially true when the program becomes more complicated by the use of the decision statements considered in this chapter and the iteration statements to be discussed in the next chapter.

In general, the steps to be taken on all programming problems are the following:

1. Analyze the problem
   Make sure the problem is completely understood. What is the required solution? What are the input data? What are the different possible cases to be considered? And so on.

2. Design the solution
   Draw a picture of the logic of the solution using flow charts. Write a summary of the solution process using both English and mathematical notation. This latter step is sometimes accomplished by *pseudocode*, which is a mathematical and English description of what is to be done, sometimes written in the form of a simple programming language.

3. Write the program
   Only when the problem has been analyzed and a solution worked out and written in terms of flow charts and/or pseudocode should you begin writing the actual program in C++. If you have outlined the solution correctly, the programming should be relatively simple.

4. Test the program
   This was discussed in Section 2.5.

If you follow these steps, it is likely that you will be able to write programs with fewer errors and save yourself a great deal of time finding and correcting errors. In addition to the above general framework (Analyze - Design - Program - Test), there are a number of useful practices to follow that are called the principles of *structured programming*.

Structured Programming

Beginning in the 1960s, there has been increasing concern about the difficulties encountered in writing correct programs in a timely fashion. There is special concern about the difficulties in modifying existing programs, especially if the people doing the modifications are not the same people who wrote the programs originally. This is true because in many cases the program may be almost impossible to understand. (Some programs became known as "spaghetti code," since the flow of control is so entangled.) Out of these concerns has emerged a consensus about programming practices that would help to remedy these problems. These practices include the following:

- *Programs should be modularized.* Functions, to be discussed in Chapter 5, are one of the tools for this in C++. Classes, to be discussed in Part III, are the primary tool.

- *The flow of control should be clear and simple.* C++ has an instruction that transfers control to any desired statement. It has been the opinion of most of the computer science community for almost 30 years that the use of such statements should be minimized, especially when the transfer is to a "far away" statement, because it makes the program harder to understand.

- *Clear and meaningful names of variables and other identifiers should be chosen as much as possible.* Cryptic variable names that have no relation to the problem should be avoided. Conflicts with standard mathematical notation should be avoided.

- *Use of programming "tricks" should be avoided.* Even though some tricks may be very clever and lead to better program efficiency, they can be counterproductive if they obscure understanding of the program. They may also become invalid if the program is modified.

- *The program should be well documented.* Programs should begin with a beginning block of comment statements describing the purpose of the program, the input necessary, the output, and so on. It is also good practice to give the author's name and date and a *modification log* that would include the author, date, and purpose of any change made to the program. Additional comment statements should be used throughout the program to describe the purpose of the key parts.

Style

Within the confines of these principles and C++ syntax, there is also the question of good *style*. Just as in writing English, each person tends to develop a style of writing programs. And, just as in writing English, one person's style may lead to programs that have a pleasing appearance and are (relatively) easy to read, whereas another person's style may lead to the opposite. Clearly, you wish to strive for good style in your writing.

Long versus Short Identifiers

One aspect of style is proper choice of identifiers. The current standard in computer science is that identifiers should be as descriptive as possible of the object being identified. Thus, for example, instead of an identifier t for a variable that is temperature, the identifier `temperature` should be used. With proper choice of identifiers a program can be made almost self-documenting, with very few additional comment statements necessary. On the other hand, standard mathematical notation typically uses only single letters; for example, f(x) denotes f as a function of x, A denotes a matrix, and so on. Since most of the applications in this book address standard mathematical problems such as integration, differential equations, and linear equations, we have tended to use as identifiers the usual mathematical notation of such problems.

MAIN POINTS OF CHAPTER 3

- The `if` statement is used for decisions. It has three basic forms:
  1. `if(logical expression)` statement
  2. `if(logical expression)`
     block of statements
  3. `if(logical expression)`
     block of statements
     `else`
     block of statements

- The mathematical relational operators $=$, $\leq$, $\geq$, $<$, $>$, $\neq$ are written in the form $==$, $<=$, $>=$, $<$, $>$, $!=$.

- Logical expressions may be formed by using the logical operators `&&` (AND), `||` (OR), and `!` (NOT) in combination with the comparison operators $==$, $<=$, $>=$, $>$, $<$, and $!=$. `if` statements may use arbitrary logical expressions.

- `if` statements may be nested, and the `else if` form of writing nested `if` statements may make them easier to read. The selection of one of several courses of action may also be implemented by the `switch` construct, in certain situations.

- Adequate planning should be done before writing a program. The use of structured programming principles will help to write correct programs.

EXERCISES

3.1 State whether and why the following program segments will print r, z, or nothing for the values r = 10.0, z = 2.0.

  a. `if((r * z) >= 2.0 )`
  `    cout << r;`
  `  else`
  `    cout << z;`

b. `if(r == z) cout << r;`

c. `if((r + z) >= 14.0)`
         `cout << r;`

3.2 Write C++ statements to carry out the following flow chart.

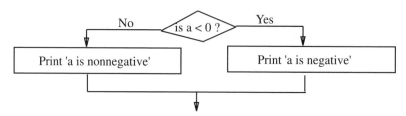

3.3 If `a = 1.0` and `b = 3.0`, what are the values of `a` and `b` after the following statements are executed?

a. `if(a > b)`
         `a = 2.0 * a;`
       `else`
         `b = 2.0 * b;`

b. `if(a >= 1.0) b = 1;`

3.4 Add to the volume program of Example 2.1 the necessary statements to test if the radius is negative. If so, print a suitable message and bypass the computation of the volume.

3.5 Are the following logical expressions true or false when `i = 0` and `a = 10.5`?

a. `(i <= 0) ||(a > 12.0)`

b. `!(i == 0)`

c. `i == 0 && a > 10.0 ||a < 9.0`

3.6 Show that Table 3.2 is equivalent to the following "truth table".

p	q	!p	p && q	p || q
F	F	T	F	F
F	T	T	F	T
T	F	F	F	T
T	T	F	T	T

3.7 Give the result of the following nested `if` statement for the cases `i = -3`, `i = 0`, `i = 21`, or `i = 7`.

```
if(i == 7)
    a = 0;
else if(i == 0)
    b = 0;
else if(i == 21)
    c = 0;
else
    d = 0;
```

Rewrite this nested if statement using the switch construct. Also, rewrite the nested if statement in the conventional indented form as in Example 3.4(a) and comment on its readability.

3.8 If a = 1, b = 2, and c = 3, what are the values of a, b, and c at the end of the following program segment?

```
if(a <= b)
    if(c > 2)
        c = 2;
if(c < 3)
    a = 0;
else
    b = 0;
```

Write the first nested if statement in a different way.

4

AROUND AND AROUND: ITERATION

In the previous chapter, we introduced the `if` statement for making decisions. The other major way to control the flow of a program is by *iteration* (or *repetition*) statements, which allow parts of a program to be repeated. The two C++ constructs to achieve iteration are the `for` statement and the `while` statement.

4.1 THE `for` STATEMENT

The program of Example 3.2 allows the input of only a single value of x. Once the corresponding function has been computed, and the output accomplished, the program then terminates. If several values are to be computed for different values of x, a separate run of the program is needed for each case. It would be much more efficient if all desired values of x could be entered, and corresponding function values computed, on each run. One way to accomplish this would be to replicate the `cin` and computational statements of the program. For more than two or three values of x, this would be very tedious. A much better way is by means of a `for` or `while` statement, each of which allows repetition of a part of a program.

Consider the following task:

$$\text{For } i = 1, 2, \ldots, n, \text{ compute and display } i^2. \tag{4.1}$$

`for` *Loops*

This can be accomplished by the program fragment shown in Example 4.1, which is called a *for loop*. It is assumed that the variables in Example 4.1 have previously been declared to be integer and that *n* has been given some specific value before the loop begins.

EXAMPLE 4.1
A for Loop

```
for(i = 1; i <= n; i++){
    i_squared = i * i;
    cout << "i squared is " << i_squared << endl;
}
```

i++

The first statement in Example 4.1 says that the following block of statements enclosed by braces is to be repeated for an initial value of i = 1 until i = n. i++ indicates that i is to be increased by 1 each time. That is,

i++   same as   i = i + 1

On occasion, we will also use

i--   same as   i = i - 1

It is considered good practice to indent the statements within the for loop, as is done in Example 4.1, and as we did with the if statement. It is also common practice to declare i within the for statement:

for(int i = 1; i <= n; i++)

if it has not previously been declared. This, however, can be risky, as will be discussed in some detail under Scope in Section 5.4.

One conventional placement of the braces in a for loop is shown in Example 4.1 and follows the same style as in Example 3.3 for an if statement. If the body of the for loop consists of a single statement, then no braces are needed, as in:

for *Loop with Single Statement*

```
for(i = 1; i <= n; i++)
   cout << i << endl;
```

Note that no semicolon follows the first statement of a for loop:

```
for(i = 1; i <= n; i++); //error
   cout << i << endl;
```

This is not a syntax error. It will compile and run, but the cout statement will be executed only once. (The technical explanation of this is that the semicolon indicates the end of the statement to be executed by the for loop. Since this statement contains nothing but the semicolon, it is a "void" statement that does nothing. But it is executed *n* times, and then the following cout statement is treated as the next statement in the program and executed just once.)

EXAMPLE 4.2
A Program for a Discontinuous Function

```
//This program computes n values of a discontinuous function
#include <iostream.h>
void main(){
  float x,f; //define x and f
  int n; //define n
  cout << "how many values?\ n"; //prompt for read of n
  cin >> n; //read n
  for(int i = 1; i <= n; i++) { //begin the for loop
    cout << "next x?\n"; //prompt and read an x
    cin >> x;
    if(x < 0) //now compute the function value
      f = x * x;
    else
      f = x + 2.0;
    //output the values of x and f
    cout << "THE FUNCTION VALUE IS " << f
         << " FOR x="<< x << endl;
  } //The end brace of the for loop
  cout << "The program is done";
}
```

As a more substantial example of a `for` loop, which combines a `for` loop with the `if` statement of the previous section, we give in Example 4.2 a program to compute values of the discontinuous function (3.2) used in Example 3.2. In this program, the number n of function values to be computed is first read and then n determines the number of times the following `for` loop is repeated. Within the `for` loop, a value of x is read, the corresponding function value is computed and then the function value as well as the value x is displayed on the terminal. The output will look like

```
THE FUNCTION VALUE IS 0.0529 FOR x=-0.23
THE FUNCTION VALUE IS 7.3 FOR x=5.3
```

4.2 THE `while` STATEMENT

A `for` loop such as in Example 4.1 or 4.2 is typically used to repeat one or more statements a specific number of times. However, if the number of repetitions is not known in advance and depends on the computation being done, then another mechanism is needed. Consider the flow chart in Figure 4.1, in which it is desired to decrease a variable *s* by an amount *t* until *s* is no longer positive. The flow chart in Figure 4.1 may be implemented by the code shown in Example 4.3, in which the statements following the `while` statement are repeated as long as `s > 0`.

Whereas the `for` loops of Examples 4.1 and 4.2 will only be executed a specified number of times, the construction of Example 4.3 can lead to an *infinite loop*, one

Infinite Loops   that will never terminate. For example, it is assumed in the program segment of Example 4.3 that both s and t are positive. However, if t were negative, the statement

Figure 4.1 A Flow Chart

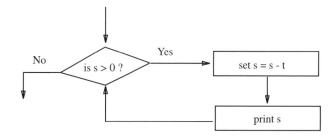

s = s - t would keep increasing s; thus s would always be positive, and the loop would never terminate. (Actually, in floating point arithmetic, if t were very large in magnitude, at some point overflow would occur and the loop would terminate with a run-time error. However, if the magnitude of t is not large, eventually the computed value of s + t would equal s because there are only a fixed number of digits to hold the sum; thereafter, s would not change and the loop would be infinite.)

EXAMPLE 4.3
A Pre-test while Loop

```
while(s > 0){
    s = s - t;
    cout << s << endl;
}
```

An alternative form of a while loop is shown in Example 4.4. This is equivalent to the loop of Example 4.3 provided that s > 0 initially. However, the statements in braces in Example 4.4 are executed before the test is made; so they are executed at least once, even if s ≤ 0 initially. The loop of Example 4.3 is sometimes called a *pretest*, whereas that of Example 4.3 is a *post-test*. The do . . . while form is usually used only if a loop must always be executed at least once. Examples 4.3 and 4.4 illustrate one conventional placement of braces in while statements.

Do *Loops*

EXAMPLE 4.4
A Post-Test while Loop

```
do{
    s = s - t;
    cout << s << endl;
}
while(s > 0);
```

We note that a for loop can always be converted to a while loop. For example, the loop of Example 4.1 could also be written as

for *Loops as*
while *Loops*

```
int i = 1;
while(i <= n){
  i_squared = i * i;
  cout << "i_squared is " << i_squared << endl;
  i++;
}
```

Here, i must be initialized and incremented in separate statements.

With both for and while loops, care must be taken to initialize variables in the loop correctly. For example, in the previous example, if the initialization int i = 1; were put after the while statement, then i would be reinitialized each time through the loop, making the loop infinite.

General logical expressions may also be used in conjunction with the while construct. The general form is

```
while (logical expression){
    statements
}
```

and an example is given in Example 4.5.

EXAMPLE 4.5
A while Loop

```
int a = 10, b = -5;
while(a > 0 && b < 0){
  a = a + b;
  b = b + 1;
}
```

Sentinels

Another common way to terminate while loops is by the use of a special number called a *sentinel*. Suppose that in the context of the volume program of Example 2.1, we wish to keep reading radii r and computing volumes v an arbitrary number of times, until r = -1 is read. This special value of r is the sentinel. We can implement this by the code fragment shown in Example 4.6.

EXAMPLE 4.6
Termination by a Sentinel

```
cout << "what is r?\n";
cin >> r;
while(r != -1){ //continue if r is not -1
  v = (4.0 * pi * r * r * r)/3.0;
  cout << r << v << endl;
  cout << "enter r or -1 to terminate \n";
  cin >> r;
}
```

Alternatively, rather than testing for a specific value of r, we could test for any negative value. Then the test in the `while` statement would be changed to r >= 0.

4.3 AN APPROXIMATION TO e^x

We next give a more important example of the use of a `while` loop. The infinite series for the exponential function is

$$e^x = \sum_{k=0}^{\infty} \frac{x^k}{k!} = 1 + x + \frac{x^2}{2} + \cdots + \frac{x^n}{n!} + \cdots$$

Suppose that we wish to approximate e^x by computing the partial sums[1]

$$E_n = 1 + x + \frac{x^2}{2} + \cdots + \frac{x^n}{n!} \tag{4.2}$$

How do we know how many terms to take? One criterion is to keep adding the terms of the series until the last one we add is suitably small:

$$\frac{|x^n|}{n!} \le \varepsilon \tag{4.3}$$

where ε is a given small parameter (for example, $\varepsilon = 10^{-5}$). Thus n in (4.2) is the smallest value for which (4.3) holds.

Example 4.7 gives a program for computing the approximation (4.2). In this program, x and eps are first read and displayed. (Recall that there must be a space or a return between the values of x and eps when you type them at the keyboard.) E is the current approximation to e^x, initialized as 1.0, and NextTerm is the next term of the series, also initialized as 1.0. Thus in the first time through the `while` loop,

```
NextTerm = (x * 1.0)/1.0 = x
```

which is added to E = 1.0. The next time through the loop

```
NextTerm = (x * x)/2.0
```

and so on. In the test in the `while` statement, `fabs(NextTerm)` is the absolute value of the current value of NextTerm; thus we repeat the statements in the `while` loop until (4.3) is satisfied. The use of `fabs` requires the directive #include <math.h> at the beginning of the program; the reason for this will be discussed in the next chapter. Each time through the `while` loop we display the current approximation E, and the corresponding value of n.

---

[1]This is not the best way to approximate e^x. It is given only as an illustration of an approximation of an infinite series.

EXAMPLE 4.7
A Program for Approximating the Exponential

```
//This program approximates an exponential by a truncated series
#include <iostream.h>
#include <math.h> //to use fabs
void main(){
  float x, eps, E = 1.0, NextTerm = 1.0, n = 1.0;
  cout << "what is x? what is eps?\n";
  cin >> x >> eps; //read x and eps
  cout << "x = " << x << " eps = "<< eps << endl; //echo
  //now do the loop for the series
  while(fabs(NextTerm) > eps){ //fabs is absolute value
    NextTerm = (x * NextTerm)/n; //compute next term in series
    E = E + NextTerm; //update approximation
    //output current approximation
    cout << "E = " << E << " n = " << n << endl;
    n++; //increase n by 1
  } //end of while loop
}
```

We give in Table 4.1 sample output for $x = 1$ with $\varepsilon = 0.0001$. As expected, the successive approximations in Table 4.1 give better and better approximations to $e = 2.718282$. (We will discuss in Chapter 7 how this output can be aligned better.)

TABLE 4.1
Output of Exponential Program of Example 4.7

```
x = 1 eps = 0.0001
E = 2 n = 1
E = 2.5 n = 2
E = 2.66667 n = 3
E = 2.70833 n = 4
E = 2.71667 n = 5
E = 2.71806 n = 6
E = 2.71825 n = 7
E = 2.71828 n = 8
```

4.4 ADVANCED LOOP CONTROL[†]

In this section we collect various more advanced techniques for controlling `for` and `while` loops. Most of these techniques are useful or needed only in special cases.

The Break and Continue Statements

In some situations, we may wish to leave a loop before its natural termination,

---

[†]This section may be omitted on first reading.

or bypass the rest of the statements in the current loop. For example, suppose that we have the following computation:

> For some initial value of m, compute $1/(m - i)$ for $i = 1, \ldots, n$, as long as $m - i \neq 0$.

Depending on the values of m and n, it may be that $m - i$ will become zero, in which case we wish to avoid the division. We can do this by means of a `continue` statement, as illustrated in the program fragment of Example 4.8. In this loop, the ratios `1.0/(m - i)` are computed and displayed for i = 1, ..., n unless m - i is zero. If m - i = 0, the `continue` statement causes the remaining statements of the loop to be skipped for the current value of i, but then the loop continues with the next value of i. Thus in Example 4.8 the effect of the `continue` statement is to bypass the statement that would divide by m - i when it is zero, but to do all other iterations of the loop. If `continue` is replaced by `break`, the loop is exited, and control is passed to the statement following the `for` loop. However, the use of a `break` statement in a `for` loop is not recommended, since we usually wish to execute the full range of values of the loop index.

EXAMPLE 4.8
A `for` Loop with a `continue`

```
for(i = 1; i <= n; i++){
  if(m - i == 0) continue; //exit iteration if m - i = 0
  cout << 1.0/(m - i) << endl;
}
```

The `break` and `continue` statements may also be used in a `while` loop. Example 4.9 illustrates this by rewriting the program fragment of Example 4.8 as a `while` loop with a `break`. In this case, if m - i = 0, the loop is terminated.

EXAMPLE 4.9
A `while` Loop with a `break`

```
i = 1;
while(i <= n){
  if(m - i == 0) break; //exit loop if m - i = 0
  cout << 1.0/(m - i) << endl;
  i++;
}
```

Beware of continue in a while Loop

The `continue` statement may also be used in a `while` loop, but caution is required. If the `break` in Example 4.9 were replaced by `continue`, the rest of the current iteration would be bypassed if m - i = 0. Thus the i++ statement would be bypassed, as well as the `cout` statement, so that i would not change and would have its same value during the next iteration of the loop. Then `continue` would be executed again, and the result would be a infinite loop.

Control of `for` Loops

We have so far only used basic `for` loops headed by a statement such as

$$\text{for(i = 0; i <= 10; i++)} \tag{4.4}$$

However, much more general control of the loop is possible. For example, the loop variable i in (4.4) takes on the values $i = 0, 1, 2, \ldots, 10$, but we can use only a subset of these by the construction.

$$\text{for(i = 0; i <= 10; i = i + 2)} \tag{4.5}$$

Loop Stride

Here, i is incremented by 2 each time and takes on only the values 0, 2, 4, 6, 8, 10. Here, 2 is called the *step* or *stride* of the loop variable and may be replaced by any integer. The step may also be negative; thus in

$$\text{for(i = 10; i >= 0; i = i - 2)} \tag{4.6}$$

Backward Loop
Loop Control

the values of i are 10, 8, 6, ..., 0, in that order, so that the loop "runs backward."
In general, the `for` loop control is of the form

for(*initialization; termination; modification*)

Here, *initialization* may be any expression that assigns an initial value to the loop index, for example, i = 3 * n or i = 32 - k, where k and n have values when the `for` statement is reached. Similarly, *termination* may be any expression, such as i * i - 2 * i < 14, that gives the termination condition. And *modification* may be any expression that changes the loop variable. The flow chart in Figure 4.2 shows the control flow of a `for` loop.

Figure 4.2 shows that the termination condition is tested before any statements in the `for` loop are executed. Thus a `for` loop need not execute any statements at all. For example, if the control statement is

Loop Need Not
Execute

for(i = k; i <= 10; i++)

and k > 10 when the loop is encountered, then it will be bypassed. Figure 4.2 also shows that the loop index is modified before the termination test so that the value of

Figure 4.2 Control
Flow of `for` Loop

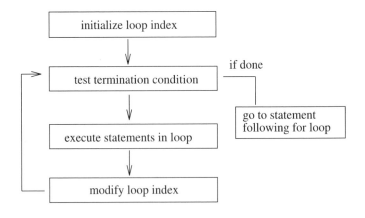

the loop index when the loop terminates is that value that terminates the loop. For example, in the loop

```
for(i = 0; i < 5; i++)
   a = a + 1;
```

Value of Index at Termination

the value of i upon termination of the loop is 5.

We note that the general `for` statement

```
for(initialization; termination; modification){
    statements
}
```

may always be converted to a `while` statement of the form

```
initialization;
while(termination){
   statements
   modification;
}
```

Bad Practices

We next discuss some legal but poor practices in the use of `for` loops. The expressions in the loop control may be null as in the examples

```
for(;;)
for(i = 0;; i++)
```

A `for` Loop May Be Infinite

In both of these cases, there is no termination criterion so that the loop must be terminated by a `break` statement. (If you forget the `break` statement, the `for` loop will be infinite!) However, the use of such a `for` loop with a `break` statement is not recommended, since it circumvents good structured programming, and the use of such loop control is not advised.

Don't Use Float Index Variables

The loop index in a `for` loop may be a floating point variable as in

$$for(float \; x = 0.1; \;\; x <= 1.0; \;\; x = x + 0.1) \tag{4.7}$$

However, this is extremely dangerous, since numbers like 0.1 do not have an exact binary representation. This, combined with rounding error, may cause the loop not to be executed the expected number of times. Consequently, *floating point variables to control a `for` loop should be used with great caution, if at all*. The effect of (4.7) may be achieved by

```
for(i = 1; i = 10; i++)
   x = 0.1 * i;
```

Another bad practice with `for` loops is to change the parameters within the loop. For example, within the body of a `for` loop begun by

```
for(i = 0; i <= n; i++)
```

Don't Change Loop
Parameters

a statement of the form `i = i + 1` that changed the loop variable `i` would result in faster termination of the loop, which may or may not be wanted. Similarly, a statement that changes n within the body of the loop would change the number of times the loop is repeated.

MAIN POINTS OF CHAPTER 4

- Repetition for a given number of times may be effected by a `for` loop.
- Repetition for an indeterminate number of times may be effected by a `while` loop. The `do while` form always executes the loop statements at least once.
- Sentinels are special values of a variable used to terminate a `while` loop.
- `for` or `while` loops may be prematurely terminated by a `break` statement or the remaining part of the current iteration bypassed by a `continue` statement.
- A `for` loop may always be rewritten as a `while` loop.
- The control statement of a `for` loop has the general form

  ```
  for(initialization; termination; modification)
  ```

  where `initialization` gives an initial value to the loop variable, `termination` gives the stopping condition, and `modification` says how the loop variable is changed.
- There are many bad but legal practices with `for` loops: using a `float` variable as the loop index; changing the loop variable or termination condition within the loop; leaving the termination condition blank.

EXERCISES

4.1 Write a program segment that inputs values a and b, adds b to a 20 times, and displays all the sums.

4.2 Write a program segment that will add b to a until the sum exceeds 100.

4.3 What are the values of r and s at the conclusion of each of the following loops, if `r = 20.0` and `s = 2.0` when the loop begins?

a.
```
for(i = 1; i <= 5; i++){
    r = r + s;
    s = s * s;
}
```

b.
```
while(s <= 100){
    r = r + s;
    s = s * s;
}
```

4.4 Run the program of Example 4.2 for several values of x, both positive and negative. Verify that the computed results are correct.

4.5 If i = 0 and g = 5, what are the values of i and g after the following program segment?

```
while((i <= 4) && (g > 0)){
    i = i + 1;
    g = g - 1.0;
}
```

4.6 Modify the program of Example 2.1 so that an integer n is first read and then the program reads n values of the radius r and computes corresponding values of the volume v. Next, modify the program along the lines of Example 4.6 so that it can compute an arbitrary number of values.

4.7 Modify the program of Example 4.7 as follows. Declare and read by a cin statement a parameter nmax, which will be the maximum number of terms allowed in the approximation. Then inside the while loop add an if statement and a break statement so as to terminate the loop if n exceeds nmax.

4.8 The series for $\cos(x)$ is

$$\cos(x) = 1 - \frac{x^2}{2} + \frac{x^4}{4!} + \cdots + \frac{(-1)^n x^{2n}}{(2n)!} + \cdots$$

Write a program similar to that of Example 4.7 to approximate $\cos(x)$ by k terms of this series. Check your program for various values of x for which you know the value of $\cos(x)$. [For example, $\cos(\frac{\pi}{4}) = \sqrt{2}/2$.]

4.9 What is the value of the int variable j at the end of the following loops if $j = 0$ when the loop begins?

   a.
```
for(int i = 8; i >= 0; i = i - 3)
    j = j + 1;
```

   b.
```
for(int i = 0; i <= 8; i = i + 2){
    j = j + 1;
    i = i + 1;
}
```

4.10 Replace the following for loop with a corresponding while loop.

```
float a = 8;
for(int i = 0; i < 10; i++){
    if(a == 0) continue;
    cout << 1/a << endl;
    a = a - 1;
}
```

chapter

5

FUNCTIONS

In many problems, it is necessary to compute a given function in different parts of the program (but perhaps with different arguments). It would be very tedious to write the statements for the same function over and over each time it was used. C++ and other programming languages permit a function to be defined once and for all, and then it may be *called* or *invoked* whenever it is needed. More generally, any program segment that is to be used repeatedly may be defined as a separate function that may be called when needed.

Using functions also has the advantage of allowing a large program to be broken up into smaller parts that may be tested separately. As discussed in Section 3.4, it is good programming practice to analyze the problem carefully and design the solution before writing any actual code. An important paradigm for this analysis and design is *top-down, stepwise refinement*, in which the overall program is first outlined as to its major parts and logical flow. Then these major parts are further broken down into subparts, and so on. Only after the problem is well understood and described in terms of a number of subtasks should the actual programming begin.

Top-Down Stepwise Refinement

The process of top-down stepwise refinement becomes more important as the size of the program grows, and many problems in science and engineering require very large programs. For example, the NASTRAN program for finite element analysis has continued to evolve and expand for almost 30 years, and now contains several hundred thousand statements. This is a large program, but it is not unusual for scientific and engineering programs to contain several thousands to millions of statements. It would be extremely difficult to write correct programs of this size without breaking them into smaller portions. This is called *modularization* of programs. Functions provide one mechanism for doing this, as they allow the possibility of breaking up a large program into smaller units, each of which may be written, compiled, and tested separately.

5.1 LIBRARY FUNCTIONS

Many of the elementary mathematical functions encountered in calculus have already been written as functions and are called *library* functions. As an example of their use, suppose that we wish to compute $\sqrt{x}$. This may be achieved by the statement

$$y = \text{sqrt}(x); \tag{5.1}$$

which sets y equal to the square root of x (x must be nonnegative or there will be a run-time error). A sample of library functions is given in Table 5.1. Recall that we used the `fabs()` function in the program of Example 4.7. [It is customary when referring to a function in C++ to include () with the name.]

TABLE 5.1
Some C++ Library Functions

Mathematics	$\sqrt{x}$	$\sin(x)$	e^x	$|x|$	x^a
C++	sqrt(x)	sin(x)	exp(x)	fabs(x)	pow(x,a)

As a second example of the use of the library functions in Table 5.1, the statement

$$y = \sin(x) + \exp(\text{fabs}(x) + x) + \text{pow}(x,\ 1.5); \tag{5.2}$$

will compute $\sin(x) + e^{|x|+x} + x^{1.5}$. Note that in this example the argument of `exp` is `fabs(x)` + x. In general, the argument of a function may be any arithmetic expression. Also, just as x must be nonnegative to compute `sqrt(x)`, so must it be nonnegative to compute `pow(x,a)` if a is not an integer. If a is an integer, however, then x may be negative.

Whenever a mathematical library function is used, the math.h library definitions
math.h *Library*   must be included by

$$\text{\#include} \ < \text{math.h}> \tag{5.3}$$

A number of other functions in the math.h library are given in Appendix 2. Recall that (5.3) was in the program of Example 4.7 so that the `fabs()` function could be used. Example 5.1 gives a simple complete program in which the statements (5.2) and (5.3) are used.

EXAMPLE 5.1
Using Library Functions

```
//This program uses some library functions
#include <math.h>
#include <iostream.h>
void main(){
  float x, y;
  cout << "Type a value of x \n";
  cin >> x;
  y = sin(x) + exp(fabs(x) + x) + pow(x,1.5);
  cout << "y = " << y;
}
```

The variable y is really not needed in Example 5.1. The last two statements could be combined into the single statement

```
cout << "y = " << (sin(x) + exp(fabs(x) + x) + pow(x,1.5));
```

5.2 USER-DEFINED FUNCTIONS

In addition to the functions that are provided by the math.h library, functions may also be defined by the programmer. Example 5.2 gives a C++ function to compute the discontinuous function

$$f(x) = \begin{cases} x^2, & \text{if } x < 0 \\ x + 2, & \text{if } x \geq 0 \end{cases} \tag{5.4}$$

used in the program of Example 4.2.

EXAMPLE 5.2
A Defined Function

```
float f(float x){
   float f_value;
   if(x < 0)
      f_value = x * x;
   else
      f_value = x + 2.0;
   return f_value;
}
```

The first statement of Example 5.2 declares f() to be a function of type float, so that values returned by f() will be floating point numbers. This statement also declares the *formal parameter* x (also called a *formal argument*, a *formal variable*, or a *dummy variable*) to be of type float. Alternatively, if x took on only integer values, and was declared as integer, f() could have been declared to be integer also by

Formal Parameter

```
int f(int x)
```

In either case, since f() takes on numerical values, it is necessary to define its type, just as if it were a variable.

In Example 5.2 the statements enclosed in braces are the *body* of the function, and define how the function value is computed. The closing brace at the end of the function body indicates that the function definition is complete. An important point is that a value is assigned to another variable, f_value, by one of the two possible assignment statements. The variable f_value then appears in the return statement signifying that f_value is returned as the function value. Note that we must declare f_value before the if statement; that is, the statement

Function Body and Return

```
float f_value = x * x;
```

within the `if` statement would be illegal. You may also see the return statement written as

return(f_value);

since `(f_value)` is the same expression as `f_value`. As opposed to some other programming languages, the function name `f` may not be used as the return variable name.

A function such as that of Example 5.2 is defined outside of `main()`. This is illustrated in Example 5.3, which is the program of Example 4.2 rewritten so as to use the function of Example 5.2.

EXAMPLE 5.3
A Program Using a Function

```
#include <iostream.h>
float f(float x); //This is the function prototype
void main(){
  float x;
  int n;
  cout << "how many x? \n";
  cin >> n;
  for(int i = 1; i <= n; i++){
   cout << "NEXT x?\n";
   cin >> x; //The next statement prints the value of f(x)
   cout << "THE FUNCTION VALUE IS " << f(x)
        << " FOR x = " << x << endl;
  }
}
  //The function f is now defined
  float f(float x){
   float f_value;
   if (x < 0)
     f_value = x * x;
   else
     f_value = x + 2.0;
   return f_value;
}
```

Function Prototype

The second statement in the program of Example 5.3 is called a *function prototype* or *declaration* statement. It defines the *function interface* and is necessary whenever a function is defined after `main()`. (When using library functions, the library `math.h` automatically provides the corresponding prototypes.) The prototype statement is just the first line of the function definition: It gives the name of the function and shows the types of the function and its parameters. (The list of parameter types is called the *signature* of the function.) Note, however, that the prototype statement is followed by a semicolon. If the function definition is put before `main()`, rather than after, no prototype statement is necessary.

Within `main()` itself, the function is called by the appearance of `f(x)` in the last `cout` statement. Note that although the variable `f_value` within the function body is set to the function value, when the function is called, the function name is used, not the variable f_value. Finally, we note that it is not legal to define a function within main() itself (or within another function), although it is permissible to have a function prototype within a function.

Once a function has been defined, it may be used just as a library function. For example, the following statements using the function of Example 5.2 may appear in different parts of a main program:

Using a Function

$$y = f(4.0); \tag{5.5a}$$

$$y = f(3.0 * w) + z ; \tag{5.5b}$$

In both cases in (5.5), the argument of the function is not x, as it is in the definition of the function in Example 5.2. The formal parameter x in the function definition is replaced by an *actual parameter* or *argument*. Thus in (5.5a) the actual parameter is 4.0, and in (5.5b) it is 3.0 * w, where the current value of w is used when the function is *invoked* or *called*, as is being done in (5.5). The function name, however, *Actual Parameters* must always be exactly what it is in the definition; thus f appears in both statements of (5.5). If f() is the function of Example 5.2, the first statement, (5.5a), will set y to 6.0, and if z is 4.0, (5.5b) will set y equal to $36.0 + 4.0 = 40.0$. The statement (5.5b) illustrates that arithmetic expressions, not just single variables, may be used as arguments when the function is called.

Functions of Several Variables

Just as in mathematics, we can also define functions of several variables. Example 5.4 gives a program for the function

$$g(x, y, z) = x^2 + y^2 + z^2$$

Note that in Example 5.4 we do not use an extra result variable but simply return the value of the computation.

EXAMPLE 5.4
A Function of Several Variables

```
float g(float x, float y, float z){
  return x * x + y * y + z * z;
}
```

Prototypes

Before main() we would put a prototype of the form

```
float g(float x, float y, float z);
```

We could also write this prototype without the argument names x, y, z:

```
float g(float, float, float);
```

All that is needed in a prototype is the number and types of the arguments; thus in Example 5.3 we could have used the prototype

```
float f(float);
```

The function in Example 5.4 might be called in main() by statements such as

```
r_squared = g(x, y, z);
a = g(1.0, 5.5, 6.1) + sin(4.9);
cout << "The value of g is" << g(x1, x2, x3) << endl;
```

The Return Statement

Multiple Return
Statements

In the functions of Examples 5.2 and 5.4 the return statement was the last statement of the function. But this is not necessary. Moreover, there can be more than one return statement. This is illustrated by the following function:

```
int match(int a, int b, int c){
  if (a == b)
    return c;
  else
    return b;
}
```

In this example, if a == b, the first return statement is executed, and control is passed back to the calling function with the return value c; if a is not equal to b, the return value is b. The function of Example 5.2 could be rewritten in an analogous way.

5.3 MORE GENERAL FUNCTIONS

In mathematics, a function of one or more variables gives a unique function value for a given set of arguments, and this is reflected by the C++ functions defined in Examples 5.2 and 5.4. However, C++ functions may perform tasks that have nothing to do with functions in mathematics. In some other programming languages, subprograms of this type are called *subroutines* or *procedures*, to differentiate them from functions that conform with mathematics. C++ has no explicit subroutine construct, but functions can also play this role. As an example of this more general type of function, Example 5.5 gives a function that computes values of two variables y and z given initial values of x and y. Note that the function has no arguments; it simply uses the current values of x and y as input when it is called in main(). Note also that it has no return value; thus the function has no type, and void is put in the type position in both the function definition and its prototype. When this program is run, it will produce the line of output:

```
x = 2 y = 9 z = 106
```

EXAMPLE 5.5
A More General Function

```
#include <iostream.h>
void example_function(); //Function prototype
float x = 2.0, y = 3.0, z; //Initialize x,y and define z
void main(){
  example_function(); //This calls the function
  cout << "x = " << x << "  y = " << y << "  z = " << z << endl;
}
void example_function(){ //Function definition
  float w = x + y; //Define and initialize w
  y = (x * x) + w;
  z = (y * y) + (w * w);
}
```

*main() Is a
Function*

The program `main()` in Example 5.5 looks rather like a function without arguments, and, indeed, that is what it is. Thus every C++ program is composed of functions, the function `main()` plus any other functions that may be defined. Most authors prefer the convention of giving `main()` the type `int` and then returning the value 0. For example, the function `main()` of Example 5.5 could be written as

```
int main(){
  example_function(); //This calls the function
  cout << "x=" << x << "y =" << y << "z =" << z << endl;
  return 0;
}
```

The return of 0 is interpreted by the operating system as the successful completion of `main()`. In the future we will use `int main()` with `return 0`. We note that functions of type `void` may also have a return statement, although it is not necessary. However, a value may not be returned by a `void` function. For example, the statement `return;` is legal but `return value;` is not.

Flow of Control

Whenever a function is called, control is transferred to that function. When the function is finished, control is then passed back to the calling function. This is illustrated schematically in Figure 5.1, where the calling function is assumed to be `main()`. The interpretation of this figure is as follows. The dots in `main()` stand for statements that are executed prior to and between calls to a function. When `function 1` is called, control is passed to that function, which executes some statements, again denoted by dots, and then passes control back to `main()`, at the point where the function was called. The same process ensues for the calls to the other functions.

Figure 5.1 Control Flow with Functions

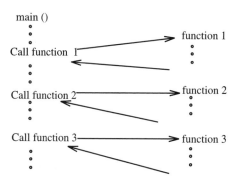

Advantages of Functions

Reusable Software

The use of functions has several advantages. The first is that additional programming may be avoided. For example, if the function of Example 5.3 were needed at several different positions in main(), the alternative to a function is to write the statements of the body of the function each time they are used. More generally, functions lead to the possibility of *reusable* software, of which the library functions are a good example. Clearly, it would be very inefficient for everyone to start from scratch to write programs for the trigonometric functions, square root, etc. Similarly, a function that you write for some particular purpose might be usable in a different program by yourself or someone else. And, as discussed in the introduction of this chapter, functions are one mechanism for the modularization of large programs.

5.4 LOCAL VERSUS GLOBAL VARIABLES

Local Variables

A variable defined within a function is *local* to that function and is usable only within that function. Thus the variable w in example_function() in Example 5.5 is local to example_function(), and main() does not know anything about it. Any formal parameters of a function are also local variables of that function. Thus in Example 5.4, x, y, and z are local variables, as is x in the function of Example 5.3.

Now suppose that in Example 5.5 we had put the statement

$$\texttt{float} \quad \texttt{x = 2.0,} \quad \texttt{y = 3.0,} \quad \texttt{z;} \qquad (5.6)$$

Global Variables

in the body of the function main() instead of outside main(). Then x, y, and z would have been local to main() and would not have been known to the function example_function(). Hence, the program would not work; indeed, there would be a compilation error saying that x, y, and z were undefined in example_function(). On the other hand, if we put (5.6) before main(), as we did in Example 5.5, the variables x, y, and z become *global*; that is, they are known to main() as well as to example_function(). In general any variable declared outside of all functions, including main(), is global, and is known to all functions following the declaration.

Usually, global variables are declared before `main()` and all other functions. Global variables are automatically initialized to zero if they are not otherwise explicitly initialized, whereas local variables are not. For example, in Example 5.5, the global variable z is initialized to zero, but the local variable w is not.

Scope

The *scope* of a variable is the set of statements for which it is defined. Thus the scope of a local variable in a function is no more than the body of the function, whereas the scope of a global variable is all the functions, including `main()`, that follow it. A global variable is said to have *global scope*.

Within a function the scope of a variable is the set of statements within any set of braces { } in which it is declared. For example, suppose that within `main()` there is the for loop

```
for(int i = 0,; i <= n; i++){
   int j = i + 1;
   cout << j << endl;
}
```

The scope of the variable j is only the body of this for loop, and it is not defined in the rest of the program. Therefore, j may be redefined in another part of the program, even as a variable of a different type. In this for loop, i is defined outside the braces of the body of the for loop, but its scope is still that of the loop, the same as j. (In older versions of C++, i was considered to be outside the scope of the for loop and so was still defined in statements following the loop.) Some people prefer to declare all variables local to a block at the beginning of the block, although others prefer to declare variables at the time they are first used.

EXAMPLE 5.6
Scope of Variables

```
int i = 2; //i is a global variable
int main(){
   int j = 3; //j is local to main()
   {int k = 4; //k is local to this block
      i = j + k; //ok.  i is global
      float j = 6; //redefines j in this block
   }
   i = j + k; //error.  k is not defined
   i = j; // j = 3, not 6
}
int function(int k){ //k is local to this function
   int j = 5; //this j is local to this function
   int i = 3; //this i is local to this function
}
```

In general, there are four types of scope: global, local in a function (function scope), local to a block within a function, and class scope, which will be discussed in Part III. We illustrate in Example 5.6 the first three different types of scope; the code in Example 5.6 is not meant to be complete. Note that the global variable i is redefined within the function. If i was not redefined, its use within the function would be that of the global variable.

MAIN POINTS OF CHAPTER 5

- C++ provides a number of functions for mathematical operations such as $\sqrt{x}$ and $\sin(x)$ in the library `math.h`.
- Additional functions may be defined by the programmer. A function may have only a single output value, given in the `return` statement, or it may be more general.
- If a function is defined after `main()`, a function prototype must be given before `main()`. The `math.h` library automatically provides the prototypes of its functions.
- `main()` is also a function and may have a type and a return value. The usual convention is a type of `int` and a return value 0.
- Variables defined within the body of a function are local to that function. Variables that are defined outside `main()`, and all other functions are global. Variables defined within a block are local to that block. The scope of a variable is the set of statements for which it is defined.
- Functions enhance the possibilities of reusing software, and provide one mechanism for the modularization of programs.

EXERCISES

5.1 Write C++ functions that will evaluate the following functions.

a. $f(x) = (1 + x^2)/(1 + 3x^3)$

b. $f(x) = (e^x \sin x)^4$

c. $f(x) = \begin{cases} \sqrt{2x}, & \text{if } x \geq 0 \\ \sqrt{-x}, & \text{if } x < 0 \end{cases}$

Then write a function `main()` that calls these functions.

5.2 Modify the volume program of Example 2.1 so as to do the computation of v in a function.

5.3 Write a single function that will compute the three quantities

$$x = (y + w)^2, \quad y = (x + w)^2, \quad z = x^2 + y^2 + w^2$$

and write a function `main()` that utilizes the computed values x, y, and z.

5.4 Rewrite the program of Example 4.7 so that the approximation to e^x is computed by a function. Run your program for various values of x and eps and compare your approximations to exp(x). Conclude that the larger x is, the more terms are needed in the series to obtain specified accuracy. Do the same for the cosine series of Exercise 4.8.

5.5 Give the three lines of output of the following program:

```
#include <iostream.h>
int i = 3;
int main(){
   j = 3;
   cout << (i + j) << endl;
   {int i = 4; j = 4;
     cout << (i + j) << endl;
   }
   cout << (i + j) << endl;
   return 0;
}
```

5.6 What is the output of the following programs?

```
#include <iostream.h>
int func(int i){
   int j = 3;
   return i + j;
}
int main(){
   int j = 4;
   cout << func(j);
   return 0;
}
```

```
#include <iostream.h>
float f(float);
int main(){
   float x = 4.0;
   cout << f(x * x);
   return 0;
}
float f(float x){
   return sqrt(x);
}
```

NUMERICAL INTEGRATION

We will now illustrate the C++ constructs, as well as the general programming principles, that we have discussed so far. We will do this in the context of the numerical approximation of an integral[1]

$$\int_a^b f(x)dx.$$

This is a relatively simple but important problem that arises in many areas of science and engineering.

In calculus, we learn to integrate certain functions exactly, for example,

$$\int_a^b x^3 dx = \frac{1}{4}(b^4 - a^4).$$

But most functions in real applications cannot be integrated in this "closed form," and the integral must be approximated. Indeed, in many applications, there may not even be an explicit formula for f; rather, there may only be a table of function values at certain points in the interval, or there may be a computer program that can calculate $f(x)$ for any x in the interval.

6.1 APPROXIMATE INTEGRATION FORMULAS

An Integral Is an Area

The integral may be interpreted as the area under the curve of f from a to b, as illustrated in Figure 6.1. An approximation to the integral may be obtained by introducing points $x_1, \ldots, x_n$ between a and b, as shown in Figure 6.1, and then approximating the integral on each subinterval (x_i, x_{i+1}) by the area of an approximating rectangle or other simple geometric figure.

---

1. If the reader has not yet studied integration, just consider the problem of approximating the area under a curve, as discussed in the text.

Figure 6.1 Integral As Sum of Areas

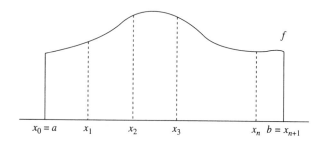

In Figure 6.2(a), the area under f from x_i to x_{i+1} is approximated by the area of the rectangle with height $f(x_i)$ and width $x_{i+1} - x_i$; thus

$$\int_{x_i}^{x_{i+1}} f(x)dx \doteq f(x_i)(x_{i+1} - x_i), \tag{6.1}$$

Rectangle Rule

which is known as the *rectangle rule*. Here, and henceforth, $\doteq$ means "approximately equal to." In Figure 6.2(b), the height of the rectangle is evaluated at the midpoint

Midpoint Rule

of the interval so that the approximation is

$$\int_{x_i}^{x_{i+1}} f(x)dx \doteq f(\frac{x_i + x_{i+1}}{2})(x_{i+1} - x_i); \tag{6.2}$$

this is the *midpoint* rule. Still another approximation is shown in Figure 6.2(c), in which the area is approximated by that of a trapezoid, rather than a rectangle. In this

Trapezoid Rule

case,

$$\int_{x_i}^{x_{i+1}} f(x)dx \doteq \frac{1}{2}[f(x_i) + f(x_{i+1})](x_{i+1} - x_i), \tag{6.3}$$

which is the *trapezoid (or trapezoidal) rule*.

We will now concentrate on the rectangle rule (6.1), but similar developments apply to the others (see Exercises 6.2 and 6.3). Using (6.1), the integral on the whole interval (a, b) is approximated by

$$\int_a^b f(x)dx = \sum_{i=0}^n \int_{x_i}^{x_{i+1}} f(x)dx \doteq \sum_{i=0}^n f(x_i)(x_{i+1} - x_i). \tag{6.4}$$

For simplicity, we will assume that the points x_i are equally spaced with spacing h;

Rectangle Rule

that is, $x_{i+1} - x_i = h$ for $i = 0, \ldots, n$. Thus (6.4) becomes

Figure 6.2 Rectangle and Trapezoid Approximations. (a) Rectangle Rule; (b) Midpoint Rule (c) Trapezoid Rule

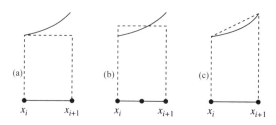

$$\int_a^b f(x)dx \doteq R(n) \equiv h \sum_{i=0}^{n} f(x_i). \tag{6.5}$$

The larger n is, the smaller h is and the more closely $R(n)$ should approximate the integral. In fact, under mild assumptions on the integrand f, it can be proved that

$$R(n) \to \int_a^b f(x)dx \quad \text{as } n \to \infty.$$

At the outset, we will usually not know how large n should be to obtain suitable accuracy in the approximation. Therefore, we will choose some reasonable $n = n_0$, and then compute the sequence of approximations

$$R_0 = R(n_0), \quad R_1 = R(2n_0), \quad R_2 = R(4n_0), \ldots, \quad R_i = R(2^i n_0), \tag{6.6}$$

doubling the number of points each time. Note that h and n are related by

$$(n+1)h = b - a; \tag{6.7}$$

thus, each time we double n, we (approximately) halve h. We continue computing the approximations R_i of (6.6) until there is very little change between successive *Convergence Test* approximations. Thus we continue until

$$|R_i - R_{i-1}| \le \varepsilon, \tag{6.8}$$

where ε is some chosen tolerance. We emphasize that because of finite precision arithmetic on a computer the parameter ε should not be chosen too small; this will be discussed further in Section 6.4.

6.2 FLOW CHARTS

The problem now is to write a C++ program to compute the sequence of approximations (6.6), where R is defined by (6.5), until (6.8) is satisfied. Before doing that, we will outline the overall structure of the computation. The program will require the integrand f, the end-points a and b of the interval of integration, the parameter ε of (6.8), and the initial number of points, n_0. The core of the computation is the evaluation of $R(n)$ by the formula (6.5). We will assume that this will be done by a function and ignore the details of this computation for now. Similarly, we will not worry about the particular function f; we assume that it will be evaluated by a C++ function. Thus we concentrate on the overall structure of the program, as shown in the flow chart of Figure 6.3.

Input Values   The first box in the flow chart shows the input information required. In the next box, $R(n_0)$ is computed and denoted by RC (for the current R). Then n is set to $2n_0$, and $R(n)$ computed and denoted by RN (for the next R). Now we begin a loop that will be repeated as long as two successive values of R differ by more than ε. Each time the loop is executed, n is doubled, RC is replaced by RN and then a new RN is computed. When convergence occurs, we output the final approximation and the value of n for which that approximation was computed. Figure 6.3 uses a

Figure 6.3 A Flow Chart for Integration

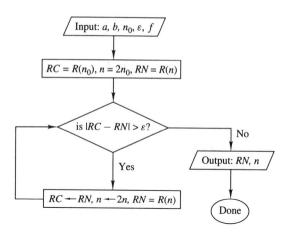

Figure 6.4 Flow Chart for Rectangle Rule Function

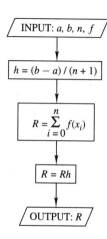

standard flow chart convention: rectangles for assignment or computation, diamonds for decisions, and parallelograms for input/output.

Next, Figure 6.4 gives a flow chart for the computation of $R(n)$. This flow chart is very simple and perhaps is not even needed; the main advantage of flow charts is to show the logical flow of the program, especially when there are decisions to be made. Note that we have still not specified the function f to be integrated; this will be furnished by a separate C++ function.

6.3 AN INTEGRATION PROGRAM

We are now ready to write the C++ program by following the flow charts of Figures 6.3 and 6.4. A function corresponding to Figure 6.4 is given in Example 6.1. The function `main()`, as well as a function for a simple integrand, is given in Example 6.2.

EXAMPLE 6.1
A Function for the Rectangle Rule

```
//This function computes the rectangle rule approximation to the
//integral of a function f over an interval (a,b) using n points
 float rectangle(float a, float b, int n){
    float h = (b - a)/(n + 1);
    float x = a;
    float value = f(x);
    for(int i = 1; i <= n; i++){
     x = x + h;
     value = value + f(x);
    }
    return value * h;
}
```

EXAMPLE 6.2
A Program for Numerical Integration

```
//////////////////////////////////////////////////////////////
// This program approximates an integral by using more and    //
// more points in the rectangle rule.  It requires a function //
// program for the integrand, the interval (a,b) of           //
// integration, an initial number of points, n0, for the      //
// approximation, and a convergence parameter, eps            //
//////////////////////////////////////////////////////////////
#include <iostream.h>
#include <math.h>
float rectangle(float a, float b, int n); //function prototype
float f(float x); //function prototype
int main(){
   int n0;
   float a, b, eps;
   cout << "a, b, n0, eps = ?\n"; //prompt for input
   cin >> a >> b >> n0 >> eps; //input a, b, n0, eps
   cout << "a = " << a << " b = " << b << endl; //echo the input
   cout << "n0 = " << n0 << " eps = " << eps << endl;
   //compute the first two approximations to integral
   float RC; //RC will hold the current value of R
   RC = rectangle(a, b, n0);
   int n = n0 + n0;
   float RN; //RN will hold the next value of R
   RN = rectangle(a, b, n);
   //repeat integral approximations until converged
   while(fabs(RN - RC) > eps){
     n = n + n; //double n
     RC = RN; //Save current approximation
     RN = rectangle(a, b, n); //Compute next approximation
     //display the results
     cout << "Integral = " << RN << " for n = " << n << endl;
   }
   return 0;
}
   //This is a particular function to be integrated
   float f(float x){
     return x/(1.0 + x * x);
   }
```

The function `rectangle()` of Example 6.1 carries out the rectangle rule. Note that the computation of h mixes floating point and integer variables; this is legitimate, and will be discussed more in Chapter 11. The `for` loop in the function computes the sum of the values of the function `f`, and then this sum is multiplied by h. An alternative formulation would be to multiply each function value by h before summing, so as to compute the area of the approximating rectangle, but this would be less efficient than a single multiplication at the end. Moreover, these extra multiplications may introduce additional error.

The function `main()` in Example 6.2 follows the flow chart of Figure 6.3 very closely. However, the flow chart does not specify how the loop is to be implemented. We have done this by a `while` loop using the `fabs` absolute value function from the `math.h` library. The `cin` statement in the main program calls for four values: a, b, no, and eps. When typing these from the keyboard, you may put spaces between each number, or use the enter key after each number. In either case, the `cin` operation will not be complete until all four numbers have been entered. Note that the first statements of the program of Example 6.2 are comments, which provide an informative introduction to the program.

The function at the end of Example 6.2 defines a particular integrand. This function can be replaced by whatever function you wish to integrate. Note that this function is called by the function `rectangle()`, not `main()`.

The program of Examples 6.1 and 6.2 produces the results shown in Table 6.1 for $a = 0$, $b = 1$, $n_0 = 2$, and $\varepsilon = 0.001$. The exact integral to 5 decimal places is

$$\int_0^1 \frac{x}{1+x^2} dx = \frac{1}{2} \log(x^2 + 1)|_0^1 = \frac{1}{2} \log(2) = 0.34657$$

so with 256 points we obtain an approximation good to almost three decimal places.

TABLE 6.1
Integral Approximations

Integral = 0.317765	for n = 8
Integral = 0.331579	for n = 16
Integral = 0.338921	for n = 32
Integral = 0.342708	for n = 64
Integral = 0.344631	for n = 128
Integral = 0.345600	for n = 256

Extensions

There are a large number of possible modifications and extensions of the programs of Examples 6.1 and 6.2; some of these are explored in the exercises of this chapter. For example, the function `rectangle()` could be replaced by another function to carry out the midpoint or trapezoid rules shown in Figure

6.2. Both of these are, in general, better than the rectangle rule. (See Section 6.4: Discretization Error.) Exercise 6.4 gives still another approximation, Simpson's rule, that is even better.

Errors

The program given in Example 6.2 has some potential problems. First, it may give completely erroneous answers; this and the general question of errors will be discussed in the next section. Second, it may not terminate if ε is chosen too small. To guard against this possibility we should permit only a prescribed number of doublings of the number of points (or a maximum number of points to be used). This will also be discussed in the next section.

6.4 DISCRETIZATION ERROR

An important source of error arises from the replacement of a "continuous" problem by a "discrete" one. For example, the integral of a continuous function requires knowledge of the integrand along the whole interval of integration, whereas a computer approximation to the integral can use values of the integrand at only finitely many points. Hence, even if the subsequent arithmetic were done exactly with no rounding errors, there would still be the error due to the discrete approximation to the integral. This type of error is called *discretization error*. All the approximations to an integral discussed in this chapter (rectangle rule, etc.) are subject to this type of error. In particular, for the rectangle rule approximation (assumed computed with no rounding error), the difference

$$E = h \sum_{i=0}^{n} f(x_i) - \int_a^b f(x)dx$$

is the discretization error.

In theory, we can estimate the error in the rectangle rule as follows. By the mean-value theorem

$$f(x_i) - f(x) = f'(z)(x_i - x),$$

where $z = z(x)$ is a point between x_i and x that depends on x. Thus the error E_i on the interval (x_i, x_{i+1}) is

Discretization Error on ith Interval

$$E_i \equiv hf(x_i) - \int_{x_i}^{x_i+h} f(x)dx = \int_{x_i}^{x_i+h} [f(x_i) - f(x)]dx$$

$$= \int_{x_i}^{x_i+h} f'(z(x))(x_i - x)dx$$

Now assume that the derivative is bounded by some constant M:

$$|f'(x)| \le M \text{ for } a \le x \le b.$$

Then

$$|E_i| \leq \int_{x_i}^{x_i+h} |f'(z(x))(x_i - x)| dx \leq M \int_{x_i}^{x_i+h} (x - x_i) dx = \frac{Mh^2}{2}.$$

Thus the total discretization error is bounded by

$$|E| \leq \sum_{i=0}^{n} |E_i| \leq \frac{n+1}{2} Mh^2 = \frac{M(b-a)h}{2} \tag{6.9}$$

since $(n+1)h = b - a$. Because M and $b - a$ are fixed quantities, this shows that the discretization error goes to zero as h goes to zero or, equivalently, as $n \to \infty$, as expected.

In the previous section, the purpose of computing approximations $R(n)$ with larger and larger n was to make the discretization error smaller. However, the larger n is, the more computation is done, and the larger the rounding error is likely to be. Thus we expect that there is a value of n for which the rounding error and discretization error are about equal, and no more progress can be made by taking n still larger. For most problems, this value of n will be considerably larger than is necessary to obtain suitable accuracy, and the discretization error will still be the dominant component of the error. However, this discussion indicates a flaw in the program of Example 6.2: If the value of ε is chosen too small (for example, 10^{-9}), the convergence test may never be passed because of the rounding error, and the program will not terminate. One way to handle this problem is as follows: In the program of Example 6.2, declare two more integer variables, say, i and imax, and read a value for imax with the cin statement. Then change the test in the while loop to

Making Sure the Loop Ends

```
while(fabs((RN - RC) > eps) && (i < imax))
```

An initialization i = 0 should come before the while loop and the increment statement i++ inside the while loop. After the while loop terminates, a subsequent test on the size of i would determine if convergence was achieved or not. It is left to Exercise 6.7 to implement these changes in the program of Example 6.2.

It is also possible for the convergence test used in the program of Example 6.2 to fail completely. Consider the function shown in Figure 6.1 and suppose we start with $n = 1$. Then the first approximation by the rectangle rule is

Convergence Test May Fail

$$\int_0^1 f(x) dx \doteq \frac{1}{2} f(0) + \frac{1}{2} f(\frac{1}{2}) = 1.$$

For the second approximation, we double n to 2, and compute

$$\int_0^1 f(x) dx \doteq \frac{1}{3}[f(0) + f(\frac{1}{3}) + f(\frac{2}{3})] = 1.$$

Since two successive approximations are equal, the convergence test is satisfied. But, as evident from Figure 6.5, the exact integral is certainly not equal to 1. This illustrates one of the many pitfalls that may occur in numerical computation.

Figure 6.5 A Difficult Integrand

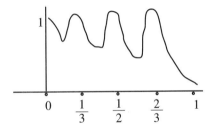

Efficiency

The discretization error bound (6.9) for the rectangle rule is of the form ch, where c is the constant $M(b - a)/2$. Such an error is denoted as $O(h)$, and we say that the error is of *order h* and the method is *first order*. A similar but slightly more complicated analysis may be done for the trapezoid and midpoint rules, and the error bound in both cases is a constant times h^2; we write this as $O(h^2)$ and say that the errors are order h^2 and that the methods are *second order*. Generally, the higher the order of the method, the more *efficient* it will be; that is, less time will be required to compute an approximation to suitable accuracy.

Another important aspect of efficiency is in knowing that certain arithmetic operations often take longer than others. Division usually takes considerably longer than multiplication or addition. Thus it is better to write `b * 0.5` than `b/2.0`. Similarly, on some machines, multiplication takes longer than addition, so that `b + b` will be more efficient than `2.0 * b`. (Good compilers will many times make optimizations like this automatically.) We will consider in many places throughout the remainder of the book the efficiency of algorithms as well as certain C++ constructs.

Summary

To summarize this chapter, we considered a relatively simple problem, numerical integration, and several possible approximations. We chose to implement the simplest one, the rectangle rule, and first constructed a flow chart showing the overall logic while ignoring the actual computations of the approximations. Based on this flow chart, and a simple flow chart for the rectangle rule approximation, we then constructed the final C++ program. This program uses most of the C++ constructs that we have discussed so far. It also illustrated, in this very simple situation, the principle of top-down refinement, in which we ignored temporarily the details of the actual coding, and only at the end did we write the function for carrying out the rectangle rule.

One further comment is necessary. One would use such a program to approximate integrals that cannot be evaluated directly. However, your program should first be extensively tested on several integrands for which the integral is easily computed.

The more tests you perform, the more confidence you will have in using the program for integrands for which you don't know the answer.

MAIN POINTS OF CHAPTER 6

- Many integrals in applications cannot be evaluated exactly and must be approximated numerically. Some simple methods for these approximations are the rectangle rule, the midpoint rule, and the trapezoid rule.

- Before attempting to write a program, the overall logic of the program should be laid out. One way of doing this is by means of flow charts.

- A program for numerical integration can consist of three parts: the function `main()` that handles input, output, and convergence test, a function that carries out the integration approximation, and a function that computes the integrand.

- Discretization error results from solving a problem such as integration by a finite discrete approximation.

EXERCISES

6.1 Run the program of Examples 6.1 and 6.2 and reproduce the results of Table 6.1. How many points, n, are required to obtain four decimal place accuracy?

6.2 The midpoint rule of Figure 6.2(b) and (6.2) leads to the approximation, corresponding to (6.6),

$$\int_a^b f(x)dx \doteq M(n) = h \sum_{i=0}^{n} f(\frac{x_i + x_{i+1}}{2})$$

Replace the function `rectangle` in Example 6.1 by a function to carry out the midpoint rule. Compute the points for which the integrand is evaluated by using only one average, and then successive additions of h.

6.3 The trapezoid approximation of Figure 6.2(c) and (6.3) leads to the approximation

$$\int_a^b f(x)dx \doteq T(n) = \frac{h}{2} \sum_{i=0}^{n} [f(x_i) + f(x_{i+1})]$$

$$= \frac{h}{2}[f(a) + f(b)] + h \sum_{i=1}^{n} f(x_i)$$

Replace the function `rectangle` in Example 6.1 by a function to carry out the trapezoid rule.

6.4   Another approximation is given by

$$\int_a^b f(x)dx \doteq \frac{h}{6} \sum_{i=0}^{n} [f(x_i) + 4f(\frac{x_i + x_{i+1}}{2}) + f(x_{i+1})],$$

which is called *Simpson's rule*. Replace the function `rectangle` in Example 6.1 by a function that carries out Simpson's Rule.

6.5   Compare the accuracy of the rectangle, midpoint, trapezoid, and Simpson's rule approximations on the integrand and parameters used for Table 6.1. Compare also the number of points, n, required to achieve convergence.

6.6   Modify the flow charts of Figures 6.3 and 6.4 if the trapezoid rule (6.3) is used instead of the rectangle rule. Do the same for Simpson's rule.

6.7   Modify the `while` statement in the integration program of Example 6.2 so as to terminate if more than `imax` doublings of n are attempted, as discussed in the text. Add a test after the `while` loop and print a suitable message if convergence has not occurred.

6.8   Insert temporary print statements in the `while` loop of Example 6.2 so that you may step through the loop examining values of RC as well as the other variables already being displayed. Then enclose these statements with `#if` and `#endif` so that they will be conditionally compiled when desired. (Be sure to add a test parameter to go with `#if`; see Conditional Compilation in Section 3.3.)

6.9   Comment out the `return` statement in the function f of Example 6.2 and add the statement `return x;` for which you can immediately compute the value of the integral for testing purposes.

6.10 Add to the `while` loop in Example 6.2 the statement `assert(n < 100)` together with the directive `#include <assert.h>`) and observe the message that this statement produces. Can you use an `assert` statement as an alternative to the method proposed in Exercise 6.7 for limiting the number of doublings of n?

chapter
7
READING AND WRITING: INPUT/OUTPUT

So far we have used only the most basic forms of input and output statements: `cin` to read data from the keyboard and `cout` to display results or messages on the monitor. Moreover, we have used the `cout` statement with only the *standard format*, and we exercised no control over the form of the output. However, there are several ways by which we may specify precisely the form we wish the output to take. In this chapter, we will discuss how to control the number of digits that are displayed by a `cout` statement as well as how to control spacing. We also discuss file input and output.

7.1 PRECISION

Consider the statement

$$\text{cout} << \text{x};\tag{7.1}$$

and suppose that `x = 37.12467`. In general, `cout` will display six figures of a number even though that much precision may not be warranted. (The number of figures displayed may vary with the compiler.) If, however, we change (7.1) to

$$\text{cout} << \text{setprecision(4)} << \text{x};\tag{7.2}$$

then x will be displayed with only four figures: 37.12. (With some compilers, 4 figures after the decimal point will be displayed, rather than a total of 4 figures.) The output will be rounded to the number of places specified so that if x had been 37.1156, 37.12 would still be printed. We note that trailing zeros are not displayed, even if significant.

Manipulators   `setprecision` is one of several *manipulators* available to use with `cout`. When you use it, or any of the other manipulators to be discussed later, you must add the directive

$$\text{\#include } \texttt{<iomanip.h>} \qquad (7.3)$$

at the beginning of your program. This makes available certain C++ functions that implement the manipulators.

Use of
`setprecision`

Other examples of the use of `setprecision` are:

$$\texttt{cout << x << setprecision(3) << y;} \qquad (7.4)$$

$$\texttt{cout << setprecision(4) << x << y;} \qquad (7.5)$$

and

$$\texttt{cout << setprecision(3) << x << setprecision(4) << y;} \qquad (7.6)$$

In (7.4), `setprecision(3)` applies only to y so that x is displayed with the standard format. Thus, if

$$x = 1.23456, \qquad y = 9.87654 \qquad (7.7)$$

then (7.4) will display 1.23456 for x and 9.88 for y. In (7.5), `setprecision(4)` applies to both x and y so that 1.234 will be printed for x and 9.876 for y. Finally, (7.6) will print 1.23 for x and 9.877 for y.

Persistence

Another situation is the following. Suppose that (7.4) has been executed and somewhat later in the program is the statement

$$\texttt{cout << x << y ;} \qquad (7.8)$$

Then both x and y in (7.8) will be printed with 3 figures since `setprecision(3)` remains in force from (7.4). In general, the number of figures specified by a `setprecision` command will be used until another `setprecision` command is encountered. We say that `setprecision` is *persistent*.

`cout.`
`precision`

An alternative way to set the precision is illustrated by the following statements:

```
cout.precision(4);
cout << x;
```

These statements have exactly the same effect as (7.2). `cout.precision` is also persistent, but it is not a manipulator and does not require the use of (7.3). [`precision()` is actually a member function of a class, as will be discussed in Part III.] Note that we cannot write

```
cout.precision(4) << x; //error
```

`cout.precision(4);` must be a separate statement.

`cout. setf`

We can also change the meaning of `setprecision` (or `cout.precision`) as follows. The statements

$$\begin{array}{l} \texttt{cout.setf(ios::fixed);} \\ \\ \texttt{cout << setprecision(4) << x;} \end{array} \qquad (7.9)$$

now interpret the 4 in `setprecision` to mean 4 digits after the decimal point, rather than 4 digits total. Thus, if again $x = 37.12467$, 37.1247 would now be displayed. Moreover, trailing zeros are also shown so that if $x = 37.120031$, 37.1200 would be displayed. `setprecision(0)` will give zero places after the decimal point. `cout.setf` is persistent and will remain in force until `cout.unsetf(ios::fixed)` is used.

7.2 SPACING

All the statements (7.4)–(7.6) will print x and y contiguous to each other, but spaces may be added between them in a number of ways. First, they may be added explicitly as in

$$\text{cout} << x << " \quad " << y; \tag{7.10}$$

where the number of spaces between the quotes will be printed. Thus, if (7.10) is executed with x and y given by (7.7), then

$$1.23456 \qquad 9.87654 \tag{7.11}$$

will appear on the monitor.

The setw
Manipulator

A usually better approach to spacing is by means of the *setwidth* manipulator, setw. The statement

$$\text{cout} << x << \text{setw}(16) << y; \tag{7.12}$$

will establish a *field width* of 16 characters for y, and y will be right justified (i.e., put to the far right) in the 16 spaces allocated by `setw(16)`. Thus, if x and y are given by (7.7), (7.12) will produce

```
1.23456        9.87654
```

There are nine blanks between the numbers since 9.87654 requires 7 spaces, including one for the decimal point. The setw manipulator is not persistent and applies only to the variable immediately following it. Thus

```
cout << setw(16) << x << y;
```

will set a field width of 16 for x, but *not* for y. Whenever setw is used, `iomanip.h` must be included by the statement (7.3).

cout. width

An alternative way to obtain the same effect as (7.12) is

```
cout << x;
cout.width(16);
cout << y;
```

In this case, as with `cout.precision()`, `cout.width(16)` must be a separate statement, and it is not necessary to include `iomanip.h`. Also, as with setw, `cout.width()` is not persistent.

The setw manipulator may also be used with integers. (setprecision, however, does not pertain to integers.) For example, if i and j are integers, the statement

```
cout << setw(6) << i << setw(6) << j;
```

will print both i and j right justified in fields of width 6.

EXAMPLE 7.1
An Output Program

```
#include <iostream.h>
#include <iomanip.h> //Need this for manipulators
int main(){
    int i = 33, j = -25;
    float x = 1.234567, y = 9.876543;
    cout << setprecision(3);
    cout << setw(10) << x << setw(7) << y << endl;
    cout << setw(7) << i << setw(7) << j << endl;
    return 0;
}
```

We give in Example 7.1 a complete program that illustrates some of the above constructions. The output from the program of Example 7.1 is

```
    1.23       9.88
   33        -25
```

with six blanks before the first number in the first line and three before the second. Since the precision has been set to three decimal figures, both numbers have been rounded to that precision. The field widths for i and j have been set so that i and j align with the integer parts of x and y.

Tabs

Still another way to set spacing, especially for tables, is by means of *tabs*, as illustrated by the statements

```
cout << "i\t\tx\n";
cout << i << "\t\t" << x << endl;
```

Here, the escape character \ is followed by t, and each \t puts in a "tab worth" of spacing (usually about 10 characters). If the variables i and x have the values 3 and 20.89 when the above statements are executed, the output will look like

```
i                   x
3                   20.89
```

with about 20 spaces between the two columns.

The setfill
Manipulator

Another manipulator, setfill, may be used in conjunction with setw in the following way:

```
cout << setfill('$') << setw(6) << "10";
```

This will produce the output

```
$$$$10
```

in which the character specified in setfill, $ in this case, fills up the remainder of the field specified by setw. Any character may be used in this context.

7.3 FILE INPUT/OUTPUT

All the programs so far in this book have displayed their output on the monitor screen, but you may want your output to be put in a file on a disk. Example 7.2 gives a modification of Example 2.1, the program that computed the volume of a sphere, in which the results are written to a file as well as displayed on the monitor.

EXAMPLE 7.2
A Program with File Output

```
//This program computes the volume of a sphere
#include <fstream.h> //for file output
#include <iostream.h>
int main(){
  float r,v;
  const float pi = 3.141593;
  cout << "r = " << flush; //Don't set for new line
  cin >> r;
  v = (4.0 * pi * r * r * r)/3.0;
  cout << "My Output is \n";
  cout << "r = " << r << " v = " << v << endl;
  ofstream prt("OUT"); //Define a file named OUT
  prt << "My Output is\n"; //Put heading in file
  prt << "r = " << r << " v = " << v << endl; //Output to file
  return 0;
}
```

A *file* consists of a collection of numbers or alphabetic characters, and is labeled by a *file name*. The statements

$$\text{ofstream prt("OUT");} \tag{7.13a}$$

File Output

$$\text{prt < < "My Output Is\textbackslash n";} \tag{7.13b}$$

$$\text{prt << "r = " << r << " v = " << v << endl;} \tag{7.13c}$$

in Example 7.2 will set up a file named OUT containing My Output Is and the labeled r and v values. Note that the statements (7.13b,c) are exactly the same as the corresponding cout statements in Example 7.2 except that cout has been replaced by prt. This is common practice; you will see on the monitor screen exactly the same thing that is being written to the file. In addition to (7.13), the directive

Include Directive for fstream.h

$$\text{\#include <fstream.h>} \tag{7.14}$$

is also needed.

The use of only the filename OUT in (7.13a) assumes that this is a file in the current directory on the current disk drive. To make sure that the file is stored where you desire, it may be prudent to give the full path of the file, as shown in PC style by

Problem with Full Path

$$\text{ofstream prt("A:\textbackslash PROGRAMS\textbackslash OUT");   //Problem}$$

This says that the file OUT should be in the directory PROGRAMS on disk drive A. But a problem arises when using directories in this fashion because \ is interpreted as the escape character. To achieve the desired effect, we must use another \ as an escape character, as discussed before in other contexts. Thus we would write

Use Escape Character

```
ofstream prt("A:\\PROGRAMS\\OUT"); //ok
```

Here the first \ before and after PROGRAMS is the escape character so that the next \ is interpreted in the correct way.

All the manipulators of Sections 7.1 and 7.2 (setprecision, setw, etc.) may also be used in writing to a file. For example, the output statement in Example 7.2 could have been written as

Can Use Manipulators with File Output

```
prt << setprecision(4) << " r = " << setw(8) << r
    << " v = " << setw(8) << v << endl;
```

The statement ofstream prt("OUT"); should only be executed one time if you wish to continue writing additional data to a file. For example, consider the loop:

```
for(int i = i; i <= 4; i++){
  cin >> r;
  ofstream prt("OUT"); //problem
  prt << " r = " << r << endl;
}
```

Here, the statement ofstream prt("OUT"); reinitializes the connection to the file OUT each time through the loop. The effect is that prior data sent to OUT is lost, and at the termination of the loop only the last value of r is in the file.

File Input

Just as we may send output to a file, we may also read input from a file. The commands are similar to those of (7.13), for example:

$$\text{ifstream read("data");} \qquad (7.15a)$$

$$\text{read >> r1 >> r2;} \qquad (7.15b)$$

File Input

Here, it is assumed that you wish to read from a file named data, so that this file name is given in (7.15a). Then (7.15b) will read two items from the file data and assign them to the variables r1 and r2. The directive (7.14) must also be used. Just as with cin, the values of r on the file should be separated by spaces or returns.

The words prt in (7.13) and read in (7.15) are arbitrary identifiers, and many programmers prefer to use fout and fin instead (for file out and file in) or other identifiers. Whatever identifiers you choose are defined by the statements (7.13a) and (7.15a).

Reading to the End of a File

The statement (7.15b) reads exactly two items from the file `data`, but in some situations you may wish to continue reading from a file until there are no more entries to be read. One common way to achieve this is illustrated by the following statements, in which it is assumed that `read` has been defined by (7.15a):

```
while(read >> r){
  cout << r << endl;
}
```

In order to understand this rather strange construction, recall that a `while` statement tests its logical expression for zero (false) or nonzero (true). The normal value of the expression `read >> r` is a reference to the input source stream and will be nonzero; hence, the next item of data on the file will be read, and the `while` loop will continue. However, if the source stream has been exhausted (there are no more data in the file), then `read >> r` is set to zero, and this terminates the `while` loop.

The other common way to terminate a read at the end of the file is illustrated by

```
read >> r;
while(!read.eof()){
  cout << r << endl;
  read >> r;
}
```

Here `read.eof()` is nonzero if there are no more items in the file and zero if there are. Hence, `!read.eof()` is interpreted as true if there are still items to be read, and false if there are no more items. In the latter case, the loop is then terminated. Note that data should be read before the loop begins so that the first test for the end of file is meaningful.

7.4 THE `printf()` FUNCTION†

Another way to print is with a function called `printf()`. This function originated in C, but is available in C++ in the `stdio.h` library. We do not recommend its use, but you are likely to see it in some programs. Here is an example.

If `x = 1.23456` and `y = 9.87654`, the statement

$$printf("x = \%.2f \quad y = \%.4f\backslash n", x, y) \tag{7.16}$$

will print

$$x = 1.23 \quad y = 9.8765 \tag{7.17}$$

The `f` in the argument `%.2f` of (7.16) is called a *conversion character*, and indicates that a floating print number is to be printed. If an integer were to be printed, the

---

†This section may be omitted without loss of continuity.

conversion character would be d instead of f. The .2 in the argument specifies that x is to be printed with two figures after the decimal point and the .4 before the second f specifies that y is to be printed with four figures after the decimal point.

The information between the quotes in (7.16) is called the *control string* or *format string*. x = and y = are characters to be printed as is. The % indicates that conversion characters follow; these are not printed but describe how the data are to be printed, as discussed above. The three blanks shown in (7.17) are a consequence of the three blanks before y in the control string. Following the control string are the names of the variables to be printed. As with cout, the \n in the control string calls for movement to a new line after the data are printed. If \n were placed before y, there would be spacing to a new line before y is printed.

An alternative way to obtain the spacing in (7.17) is by the statement

```
printf("%2s%.2f%5s%.4f \n," "x =", x, "y=", y);
```

Here, 2s indicates that x = is to be printed in a field width of two, and 5s indicates that y = is to be printed with a field width of five; thus three spaces will be to the left of y =. s is a conversion character, and therefore 2s and 5s are preceded by %. Note that in this construction, the character strings x = and y = are listed as entities to be printed, along with the values of x and y. In general, any number of spaces may be specified in this fashion, which may be preferable to explicitly leaving spaces, as in (7.16). In particular, if the 2s is changed to 20s, there would be 20 spaces before x = is printed, thus centering the output.

Another way to obtain spacing is to specify a field width with the f (or d) conversion characters. For example,

```
%6.2f
```

specifies a field width of 6. Thus, if the number to be printed is 2.413, it will be printed as

```
bb2.41
```

where the b's indicate leading spaces. In general,

```
%m.nf
```

specifies a field width of m with n digits to the right of the decimal point. If m is larger than the total number of decimal digits to be printed, including n digits to the right of the decimal point, then any remaining space in the field will be used for leading blanks. Note that if m is specified to be too small to hold the number to be printed, enough space will still be allocated to print the number. Field widths may also be used with integers. Thus

```
%14d
```

will set a field width of 14.

Whenever the function printf() is used, we must also use the directive

```
#include <stdio.h>
```

but the libraries `iostream.h` and `iomanip.h` are not needed. This is illustrated by the program in Example 7.3. This program will produce the two lines of output

```
x = 1.23   y = 9.8765
i = 123   j = 987
```

EXAMPLE 7.3
Output Using `printf()`

```
//This program prints using printf()
#include <stdio.h>
int main(){
    int i = 123, j = 987;
    float x = 1.234567, y = 9.876543;
    printf("x = %.2f    y = %.4f\n", x, y);
    printf("i = %d    j = %d\n", i, j);
    return 0;
}
```

MAIN POINTS OF CHAPTER 7

- The `setprecision` and `setw` manipulators allow customized output by specifying the number of figures to be printed, and spacing. The library `iomanip.h` must be used with these manipulators.

- `setprecision` normally controls the total number of figures that are printed. An alternative is `cout.precision()`. When used with `setf`, these control the number of places to the right of the decimal point. Both `setprecision` and `cout.precision()` are persistent: They stay in force until changed.

- `setw` defines a field width and the following output is right justified in the field. `cout.width()` does the same. Neither is persistent.

- Output to a file may be achieved by using `ofstream`, along with the directive `#include <fstream.h>`. In a similar way data may be read from a file by using `ifstream`.

- An alternative (not recommended) to `cout` is the C function `printf()`. Rather than manipulators, `printf()` uses the `f` conversion character to specify the number of decimal places in floating point numbers, the `d` conversion character to print integers and the `s` conversion character to control spacing.

EXERCISES

7.1 Show the output of the following statements assuming that a = 41.16, b = 21.12, i = 52, and j = 101. Use _ to show blanks.

```
cout << setprecision(3) << a << setw(12) << b << endl;
cout << setw(6) << i << setw(8) << j << endl;
```

7.2 Rewrite the statements of Exercise 7.1 so that a and b appear right justified in fields of width 10, and i and j align under the integer parts of a and b.

7.3 Rewrite the statements of Exercise 7.1 so that the labels a =, b =, i =, and j = print before the respective numbers.

7.4 Use the statement of (7.9) and the setprecision and setw manipulators to rewrite the cout statement inside the while loop of Example 4.7 to modify the output shown in Table 4.1. In particular, print the values of E to four decimal places with trailing zeros shown and align the column of n values with 5 spaces between the n column and the E column. Also, send the output to a file.

7.5 Rewrite the first statement of Exercise 7.1 using cout.precision() and cout.width() so as to obtain exactly the same precision and spacing.

7.6 Write statements to print the data a = 432.123, b = 1.214, c = 0.011 with 3 figures to the right of the decimal point.

7.7 Construct a file of, say, 5 or 10 values, that will be radii of spheres. Each entry in this file should be terminated by a space or a return. Then modify the program of Example 2.1 so that it repeats the calculation of the volume of a sphere for each value of a radius read from the file, and terminates when the end of file is reached.

8

SOLUTION OF
NONLINEAR EQUATIONS

We next consider another problem that is extremely important in scientific and engineering computing: the numerical solution of a nonlinear equation

$$f(x) = 0. \tag{8.1}$$

Some special cases of this problem are to find the roots of a polynomial

$$f(x) = a_0 + a_1 x + \cdots + a_n x^n, \tag{8.2}$$

or to solve transcendental equations such as

$$x - \sin x = 4. \tag{8.3}$$

Also, in many problems, (8.1) arises as the result of wishing to minimize (or maximize) a differentiable function $g(x)$. By calculus, a necessary condition for a (local) minimum is that $g'(x) = 0$; in this case, $f(x)$ is $g'(x)$ in (8.1).

By the Fundamental Theorem of Algebra, a polynomial of degree n has exactly n roots, if multiple roots are counted appropriately. However, even if the coefficients $a_0, \ldots, a_n$ of the polynomial are all real, some of the roots may be complex; for example, the equation $x^2 + 1 = 0$ has the two solutions $\pm\sqrt{-1}$. Although in some problems it may be important to find complex solutions of (8.1), we will make the simplifying assumptions that x is real, $f(x)$ is real, and we wish only to find real solutions.

There are many known methods to approximate solutions of (8.1). In this chapter, we will describe two of the most basic: the bisection method and Newton's method. The bisection method is reliable but slow, whereas Newton's method is faster but not as reliable. After discussion of these two methods, we will show how to combine them into a method that will be better than either method separately.

8.1 THE BISECTION METHOD

If f is a continuous function with $f(a) < 0$ and $f(b) > 0$, then it is intuitively clear (and can be rigorously proved) that (8.1) has a solution between a and b. This is illustrated in Figure 8.1, where $x^\star$ is the solution.

Now let $x_1 = (a + b)/2$ be the midpoint of the interval (a, b). If $f(x_1) = 0$, we are done. If $f(x_1) \neq 0$ and $f(a)$ and $f(x_1)$ have different signs, then the root $x^\star$ is in the interval (a, x_1); otherwise, it is in (x_1, b). In either case, the root is now known to lie in an interval half the size of the original interval. We then repeat the process, always keeping the interval in which $x^\star$ is known to lie, and evaluating f at the midpoint of this interval in order to obtain the next interval. For example, a typical sequence of steps is shown in Figure 8.2 and the corresponding points x_1, x_2, and x_3 in Figure 8.1.

At each step of the bisection process we obtain a new interval containing the root, and this interval is half the size of the previous interval. Clearly, the midpoints of this sequence of intervals will converge to a root. Thus we may base a criterion *Convergence Test* for terminating the bisection procedure on the following convergence test. Given a parameter ε, say $\varepsilon = 10^{-3}$, we continue the bisection process until the current interval is of length no greater than ε. Then we take the midpoint of this interval to be the final approximation to the solution. This guarantees (in exact arithmetic) that the solution is correct to $\varepsilon/2$. Thus the parameter ε is chosen on the basis of what accuracy is desired in the final approximation.

Figure 8.1 A Continuous Function and the Bisection Method

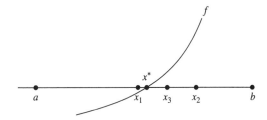

Figure 8.2 The Bisection Method

$f(x_1) < 0$. Hence, $x_1 < x^\star < b$.   Set $x_2 = \dfrac{1}{2}(x_1 + b)$

$f(x_2) > 0$. Hence, $x_1 < x^\star < x_2$.   Set $x_3 = \dfrac{1}{2}(x_1 + x_2)$

$f(x_3) > 0$. Hence, $x_1 < x^\star < x_3$.   Set $x_4 = \dfrac{1}{2}(x_1 + x_3)$

EXAMPLE 8.1

A Bisection Program

```
///////////////////////////////////////////////////////////////
// This program approximates a root between a and b of a function /
// f.  It is assumed that f(a) is negative and f(b) is positive.  /
// The program requires a function subprogram for f, the points a /
// and b, the convergence tolerance, eps, and the maximum number  /
// of bisections allowed, kmax.  The program prints the successive/
// intervals and their midpoints as the bisection proceeds.       /
///////////////////////////////////////////////////////////////
#include <iostream.h>
#include <iomanip.h>
#include <math.h> //for log x
float f(float); //function prototype
int main(){
 int kmax;
 float a, b, eps;
 cout << "input a, b, eps, kmax\n";
 cin >> a >> b >> eps >> kmax;
 cout << "The input data are\n";
 cout << "a = " << a << "    b = " << b;
 cout << "    eps =" << eps << "    kmax ="<< kmax << endl;
 cout << "The results are\n"; //display table headings
 cout << "k      x            f(x)        a        b\n";
 int k = 1;
 cout.setf(ios :: fixed);
 float x = 0.5 * (a + b); //first bisection
 while((k <= kmax) && ((b -a) > eps)){
  if(f(x) < 0)   //choose
    a = x;       // the
  else           // right
    b = x;     // interval
  x = 0.5 * (a + b); //next bisection
  //print k, x, y, a, b
  cout << k << setprecision(6) << setw(10) << x << setw(10) << f(x)
       << setprecision(4) << setw(8) << a << setw(8) << b << endl;
  k++; //increment loop variable
 }
 if(k > kmax) cout << "no convergence";
 return 0;
}
//this is the function
float f(float x) return -1.0/x - log(x) + 2.0;
```

Example 8.1 gives a C++ program for the bisection method. The program reads the endpoints a and b of the interval, the convergence parameter eps and also a maximum number of bisections to be allowed, kmax. The reason for kmax will be discussed in Section 8.3. If this maximum is exceeded before convergence is achieved, the message "no convergence" is printed. The while loop performs the bisection process and displays the iteration number, the current interval, a, b, the midpoint x of this interval, and the corresponding function value f(x).

We now apply the bisection program of Example 8.1 to the function shown in Figure 8.3. This function is defined for positive values of x and has two roots, as indicated in the figure. We are attempting to approximate the one between 0 and 1. The bisection method illustrated by Figures 8.1 and 8.2 is based on the assumption that $f(a) < 0$ and $f(b) > 0$. The method works just as well if $f(a) > 0$ and $f(b) < 0$, but the decisions must be reversed from those in Figure 8.2. For example, if $f(x_1) < 0$, then $a < x^* < x_1$, and so on (see Exercise 8.1). Alternatively, when $f(a) > 0 > f(b)$, we may work with $-f(x)$ and use the original algorithm. This is what we have done for the function of Figure 8.3, using its negative in the function in Example 8.1.

Table 8.1 gives sample output for the data a = 0, b = 1, eps = 0.01, kmax = 10. Note that the function is not defined for 0, but a = 0 is an acceptable value since the bisection method will never evaluate the function at the endpoints of the

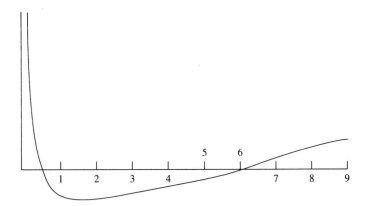

Figure 8.3 The Function $f(x) = 1/x + \ell n(x) - 2$

TABLE 8.1
Output from Bisection Program: Root = 0.31784443

The input data are
a = 0   b = 1   eps = 0.01   kmax = 10

The results are

k	x	f(x)	a	b
1	0.250000	-0.613706	0.0000	0.5000
2	0.375000	0.314163	0.2500	0.5000
3	0.312500	-0.036849	0.2500	0.3750
4	0.343750	0.158750	0.3125	0.3750
5	0.328125	0.066742	0.3125	0.3437
6	0.320312	0.016507	0.3125	0.3281
7	0.316406	-0.009766	0.3125	0.3203

interval. The sample output shows that the intervals are converging to the root, whose value is 0.3178. After seven bisections, we have approximated this root to almost three decimal places. Were the parameter eps smaller, still more accuracy would be obtained after further bisections (Exercise 8.2).

The program of Example 8.1 may, of course, be used for approximating roots of other functions simply by changing the function program.

8.2 NEWTON'S METHOD

The bisection method is simple and reliable, but it is rather slow. A potentially much faster method may be derived as follows. Let x_0 be an approximation to the solution x^* and draw the tangent line to the curve of f at x_0, as shown in Figure 8.4. Then we take the intersection of this tangent line with the x axis as a new approximation to x^*.

We now need to formulate this geometric procedure analytically so that we may compute the new approximation x_1. The tangent function is

$$T(x) \equiv f'(x_0)(x - x_0) + f(x_0).$$

The next approximation, x_1, is then the solution of the equation

$$T(x) = 0,$$

which gives

$$x_1 = x_0 - \frac{f(x_0)}{f'(x_0)}.$$

We may now repeat this process, replacing x_0 by x_1 to obtain another approximation, and then repeat again, and so on. Thus we can generate a sequence of approxima-

Newton's Method   tions 'by

$$x_{k+1} = x_k - \frac{f(x_k)}{f'(x_k)}, \quad k = 0, 1, \ldots \tag{8.4}$$

This is *Newton's method* for approximating a solution of a nonlinear equation.

Figure 8.4 Newton's Method

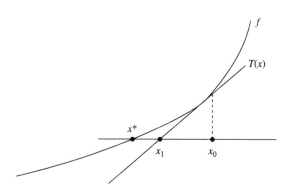

We next discuss some properties of Newton's method. First, it is clear that $f'(x_k)$ must be nonzero to avoid division by zero in (8.4). Geometrically, if $f'(x_k) = 0$, the tangent line at x_k is horizontal, and thus has no intersection with the x axis. Assuming that we avoid division by zero, the two main properties of Newton's method are:

Convergence

$$\text{If } x_0 \text{ is sufficiently close to } x^\star, \text{ then } x_k \to x^\star \text{ as } k \to \infty. \tag{8.5}$$

Quadratic convergence If f is twice continuously differentiable, if $f'(x^\star) \neq 0$, and if $x_k \to x^\star$ as $k \to \infty$, then

$$|x_{k+1} - x^\star| \doteq c|x_k - x^\star|^2 \text{ as } k \to \infty \tag{8.6}$$

where c is some constant. The second property of *quadratic convergence* ensures that once the iterates get close to the solution, they converge very rapidly, approximately doubling the number of correct digits each iteration.

As an illustration of quadratic convergence in Newton's method, we again consider the function of Figure 8.3. Table 8.2 contains a summary of the first six iterations of Newton's method using the starting value $x_0 = 0.1$; the root is obtained to 8 decimal places in 6 Newton iterations. Note that once an approximation is "close enough" (in this case, after three iterations), the number of correct digits doubles in each iteration, showing the quadratic convergence.

In order to stop the Newton iteration we will use the convergence test

$$|f(x_k)| \leq \varepsilon \tag{8.7}$$

where ε is a small number, say 10^{-4} or 10^{-6}. The criterion (8.7) does not guarantee that x_k is close to $x^\star$, but it does say that we have found an approximate solution in the sense that $f(x_k)$ is small. We next give a pseudocode for Newton's method in Figure 8.5, using the convergence test (8.7) as well as a maximum allowed number of iterations. In this pseudocode, a `while` statement terminates the Newton iteration if there is convergence or if $f'(x_i) = 0$ or if the prescribed maximum number of iterations has been exceeded. Subsequent tests are then required to ascertain which

TABLE 8.2
Convergence of Newton's Method to Root = 0.31784443

Iteration	x_{i-1}	$f(x_{i-1})$	x_i	Number of correct digits
1	0.1	5.6974149	0.16330461	0
2	0.16330461	2.3113878	0.23697659	0
3	0.23697659	0.7800322	0.29438633	1
4	0.29438633	0.1740346	0.31576121	2
5	0.31576121	0.0141811	0.31782764	4
6	0.31782764	0.0001134	0.31784443	8

Figure 8.5 A Pseudocode for Newton's Method

input: x_0, ε, max i, f, f'

set $i = 0$

while ($i \leq$ max i and $|f(x_i)| > \varepsilon$ and $f'(x_i) \neq 0$)

$\quad x_{i+1} = x_i - f(x_i)/f'(x_i)$

$\quad i = i + 1$

end while

if $f'(x_i) = 0$, output error message

else if $i >$ max i, output message

else output x_i as solution

of these conditions occurred. It is left to Exercise 8.3 to write a C++ program for Newton's method following the pseudocode of Figure 8.5.

8.3 ERRORS AND A COMBINED METHOD

Convergence Error

Iterative methods such as the bisection method or Newton's method are subject to error that is somewhat akin to discretization error. In an iterative process, a sequence of approximations to a solution is generated with the hope that the approximations will converge to the solution; in many cases mathematical proofs can be given that show convergence will occur as the number of iterations tends to infinity. However, only finitely many such approximations can ever be generated on a computer, and, therefore, we must necessarily stop short of mathematical convergence. The error caused by such finite termination of an iterative process is sometimes called *convergence error*, although there is no generally accepted terminology here.

Wrong Sign for f

There is another aspect of convergence error when rounding error is present. Consider Newton's method. When the iterates x_k approach the solution $x^\star$, $f(x_k)$ becomes small, and the sign of $f(x_k)$ may be evaluated incorrectly because of rounding error. If the sign of $f'(x_k)$ is correct, which will be true if $f'(x_k)$ is not small, then the change in the iterate, $f(x_k)/f'(x_k)$, has the wrong sign, and the computed next iterate moves in the wrong direction. The same difficulty may occur in the bisection method, since if the sign of $f(x_k)$ is incorrect, a wrong decision will be made as to the next interval to retain. Because of this, with both of these methods, as well as others, one will tend to see an erratic behavior of the iterates when errors in the sign of $f(x_k)$ occur, and then it is no longer useful to continue the iteration. This is one reason that only a maximum number of iterations should be allowed, since convergence may never occur if the convergence parameter is too small.

Bad Behavior of Newton

With the bisection method without rounding error, convergence is guaranteed once we have a suitable starting interval. With Newton's method the situation is different. The convergence property (8.5) of Newton's method stated that if x_0 is sufficiently close to the solution $x^\star$, the Newton iterates will converge to $x^\star$. Convergence will also occur if any iterate x_k gets sufficiently close to $x^\star$; otherwise,

Figure 8.6 Divergence of Newton Iterates

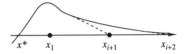

various types of bad behavior may occur. For example, $f'(x_k)$ may be zero and the iteration stops. Another type of bad behavior is illustrated in Figure 8.6, in which successive iterates diverge away from x^*.

In Figure 8.6, we wish to find the root at x^*, but the approximation x_1 is such that successive Newton iterates move away from x^*. If there is another root to the right of x^*, the Newton iterates may converge to it. Or there may be no root to the right of x^*, and the iterates diverge to infinity. This would be the case if, for example, the function behaved like e^{-x} for large x.

A Combined Newton–Bisection Method

In order to mitigate these problems with Newton's method, we will combine it with the bisection method, as indicated by the flow chart in Figure 8.7.

Newton Iterates
Can't Escape

In this figure, we assume that we start with x_0 in an interval (a, b) for which $f(a) < 0 < f(b)$. If $f'(x_0) = 0$, or if the next Newton iterate is outside the current

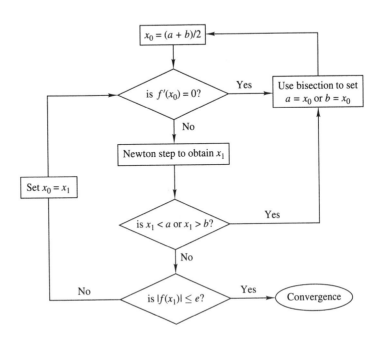

Figure 8.7 Combined Newton-Bisection Method: $f(a) < 0 < f(b)$

interval (a, b), we do a bisection step to obtain a smaller interval, also denoted by (a, b), and then try the Newton process again. Thus the Newton iterates cannot "escape." Either they converge, or we obtain a shrinking set of intervals by bisection. As with the Newton and bisection methods themselves, we should add to the flow chart a test that will cause termination if a maximum number of iterations is exceeded. Exercise 8.5 requests that you write a C++ program to implement the flow chart of Figure 8.7.

There are many possible variations on Figure 8.7. For example, after a successful Newton step we could use the Newton iterate x_1 as one endpoint of a new interval in which the root lies. This would prevent another possible difficulty with Newton's method known as *cycling*, in which we can have $x_2 = x_0$ so that no convergence is obtained, even though the iterates do not diverge.

MAIN POINTS OF CHAPTER 8

- Two simple methods for approximating solutions of an equation $f(x) = 0$ are the bisection method and Newton's method.
- The bisection method is reliable but slow. Newton's method is potentially much faster (quadratic convergence) but may not converge. A combination of the two methods ensures reliability.
- Both the bisection method and Newton's method may be affected by rounding error if the sign of the function is not evaluated correctly.

EXERCISES

8.1 Redo the program of Example 8.1 so that the bisection algorithm works for the case that $f(a) > 0$ and $f(b) < 0$.

8.2 Run the program of Example 8.1 for different values of eps and kmax. How many bisections do you need to take in order to approximate the root to 6 decimal places?

8.3 Write a C++ program for Newton's method following the pseudocode of Figure 8.5. Run the program for the function of Figure 8.3 and see if you can reproduce the results of Table 8.1.

8.4 Let $f(x) = x - \sin x$. This function has three roots: 0 and one in each of the intervals $(\pi/2, 2)$ and $(-2, -\pi/2)$. Attempt to find the positive root by both the bisection method and Newton's method, using the programs of Example 8.1 and Exercise 8.3. Experiment with different starting values for Newton's method within the interval $(\pi/2, 2)$. Graph the function and try to ascertain geometrically which starting values will give convergence.

8.5 Write a C++ program for a combined Newton–bisection method along the lines of Figure 8.7, with the loop implemented by a while construct. Add a test to terminate on a given maximum number of iterations if convergence has not occurred. Run your program on the equation of Exercise 8.4 using some initial x_0 for which Newton's method didn't converge.

9
LOTS OF VALUES: ARRAYS

In Chapter 4, we discussed how to use a `for` loop to repeat a computation many times. In the context of the volume computation of Example 2.1, suppose that we wished to compute several values of v for given values of r, and then store these r and v values for further use. Each value of r would need its own storage location; we might give these values the names r1, r2, ..., and similarly for the v values. This has the disadvantage that we must define lots of variables, and we must decide in advance just how many values of r and v we wish to be able to accommodate. This is very tedious, and a much better way to handle this is by means of *arrays*. (See Exercise 9.2.)

9.1 ARRAY DECLARATIONS

The declaration

$$\text{float a[100], b[100];} \tag{9.1}$$

defines two arrays a and b, each consisting of 100 *subscripted variables*: a[0], a[1], ..., a[99] and b[0], b[1], ..., b[99]. That is, 200 positions for floating point numbers are reserved in memory, as indicated in Figure 9.1. The elements of a are stored contiguously in memory one after another. The elements of the array b are also stored contiguously in memory but not necessarily right after the array a elements. Note that the indices of the array elements are 0, 1, ..., 99, *not* 1, ..., 100.

In declaring an array, the number of elements must be given, as in

```
float a[100];
```

or as in

```
const int max = 100;
float a[max];
```

Initialization of Arrays

Arrays May Be const

Using Array Elements

Figure 9.1 Memory Allocation for Arrays

a[0]	b[0]
a[1]	b[1]
. . .	. . .
a[99]	b[99]

This latter construction is especially advantageous if several arrays of length max are to be declared; then the array sizes may all be changed by changing the single parameter max. Note that the integer max defined in this context *must* be const. Expressions are also allowed as in

```
float a[3 * max];
```

On some compilers, the declarations of very large arrays (several hundred elements or more) in main() or in other functions may cause a run-time error. If this occurs, declare the arrays as global.

Array elements may be assigned values by initialization, much as variables. For example, the statement

```
float arr[3] = {3.0, 4.1, 6.2;}
```

defines arr to be an array of length 3 and sets the elements of the array to the values shown. Not all the elements need to be initialized; for example, the statement

```
float a[100] = {3.0, 4.1, 6.2}, b[100] = {3.1, 4.5, 7.2};
```

defines a and b to be arrays of length 100, as in (9.1), and initializes the first three elements of each array. In this case, the remaining elements of the array are initialized to zero. (But it is risky to assume that all compilers will initialize these elements to zero; so explicit initializations should be done if it is important.) If a global array is not explicitly initialized, then all elements of the array are initialized to zero. However, arrays defined within a function, including main(), are not automatically initialized. An easy way to initialize an array to zero is

```
float a[100] = {0};
```

This statement will declare an array a and set all 100 elements to zero. Arrays may also be declared as constant. Thus the declaration

```
const float arr[3] = {3.0, 4.1, 6.2};
```

defines an array whose elements cannot be changed.

Assuming that the elements of the array b have been assigned values, perhaps by initialization, the assignment statement

```
a[3] = b[3] * b[3];                          (9.2)
```

will set a[3] to the square of the current value of b[3]. But, rather than addressing some particular element of an array, as in (9.2), usually we will work with arrays in the context of for or while loops. For example, in place of (9.2), we would more likely be doing the computation for all values in the array as illustrated by the program segment in Example 9.1.

EXAMPLE 9.1
A for Loop with Arrays

```
for(int j = 0; j < 100; j++)
    a[j] = b[j] * b[j];
```

Indices May Be Expressions

As well as accessing array elements by, say, a[3] or a[i], we can use any integer expression as the index; for example:

$$a[i + 1], \ a[i * j], \ a[i * i + j * 5]$$

Generally, it is not good practice to use very complicated expressions as array indices. Integer-valued array elements or functions may also be used as indices; for example, a[k(i)] is permitted, where k is either an array (change the () to [ ]) or a function.

The declaration (9.1) reserves storage for 100 values of a and b, but all of these do not need to be used in any given instance. For example, (9.2) uses only one value of each, and the first statement in Example 9.1 could be changed to

for Loops and Arrays

```
for(int j = 0;  j < 50; j++)
```

or

```
for(int j = 0;  j < n; j++)
```
(9.3)

where $n \leq 100$. But in conjunction with (9.1), the statement

```
for(int j = 0; j < 150; j++)
```

would be incorrect since values of a[j] and b[j] are not defined for $j \geq 100$. As opposed to many other programming languages, C++ does not check, either at compile time or at run time, if access is attempted to an array element outside the defined array. Thus, if a and b are defined by (9.1), in the statement

Array Errors

```
a[100] = b[100] * b[100];
```

whatever happens to be in the storage location after the array b will be used for b[100]. The multiplication might then cause a floating point run-time error. If not, the result of the computation will be stored in the location after the array a. This might wipe out a piece of data, for example, b[0], or cause a "general protection fault" error. Therefore, when using arrays you must be extremely careful not to exceed the declared limits of the array. (We will see in Part III how to design an array class that checks the array limits.)

EXAMPLE 9.2
Program with Arrays and `for` Loops

```
//This program computes the squares of n elements in an array
#include <iostream.h>
int main(){
  float a[100],b[100]; //declare two arrays
  int j, n;
  cout << "What is n?\n"; //prompt and read n
  cin >> n;
  //Read elements of array b
  for(j = 0; j < n ; j++){
    cout << "b[" << j << "]?\n";
    cin >> b[j];
  }
  //Compute and output squares of elements
  for(j = 0; j < n ; j++){
    a[j] = b[j] * b[j];
    cout << b[j] << "   " << a[j] << endl;
  }
  return 0;
}
```

Example 9.2 gives a simple complete program in which, in the first `for` loop, values for `b[0], ..., b[n-1]` are to be entered one at a time following the prompts. Each prompt will be of the form `b[j] = ?`, where the current value of `j` will be shown. Then, after the computations, pairs of values `a[j]` and `b[j]` are displayed on the monitor, one pair per line.

9.2 ARRAYS AND FUNCTIONS

Array elements may be used as arguments of functions. For example,

$$y = sqrt(a[3])$$

will compute the square root of the fourth element of the array `a`, assumed positive. Example 9.3 illustrates that arrays themselves, not just a particular element, may also be used as arguments of a function. This function computes the sum of `m` elements of an array called `c` in the function definition. In the first function call from `main()`, the arguments are the array `a` and 100 for the value of `m`. Thus the function `array_sum()` will compute the sum of all 100 elements of `a`. In the second call, the array is now `b`, and `n` is the number of elements. (The value of `n` as well as the elements of `a` and `b` are assumed to have been set by constructions, perhaps `cin` statements, not shown.) In this case, `n` may be any positive integer not exceeding 100 and the corresponding elements of the array `b` will be summed.

EXAMPLE 9.3
A Function with an Array Argument

```
//This program computes the sum of array elements
#include <iostream.h>
float array_sum(float c[100],int m); //function prototype
int main(){
  float a[100],b[100]; //Define two arrays
  int n;
  //code to set values of n and arrays a and b here
  float suma = array_sum(a,100);
  float sumb = array_sum(b,n);
  cout << "a sum = " << suma << endl;
  cout << "b sum = " << sumb << endl;
  return 0;
}
//This defines the function array_sum
//It computes the sum of m elements of the array c
float array_sum(float c[100],int m){
  float sum = 0; //declare and initialize return value
  for (int i = 0; i < m; i++) //now compute the sum
    sum = sum + c[i];
  return sum;
}
```

In the function `main()` of Example 9.3 the function `array_sum()` is called by the statement

$$\texttt{float sumb = array\_sum(b, n);}$$

Call by Value

Call by Reference

Then the current value of the integer n is passed to the function, and a temporary memory location is created by the function to hold this value. This is termed *call by value*, and call-by-value variables are local variables of the function. With the array b, however, it would be an inefficient use of memory to create temporary storage for a copy of b within the function, especially if b is a large array. Hence, the address of the array b is passed to the function, rather than the actual elements of b. This is termed *call by reference* or *call by address*. In this case, the function accesses values of b directly from b's originally assigned memory, and no copy of the array is made within the function. Note that, however, in a statement such as

$$\texttt{x = sqrt(b[3]);}$$

where only a single element of the array is the argument, the value of b[3] would be passed to the function. A more detailed discussion of call by value and call by reference will be given in Chapter 14.

Adding Two Arrays

Example 9.4 gives another example of a function with array parameters. This function computes the n sums

$$c_i = a_i + b_i, \quad i = 1, \ldots, n, \qquad (9.4)$$

and does not return a value; rather, its output is the array c. The parameters of the function of Example 9.4 are the arrays a, b, and c, and an integer n.

EXAMPLE 9.4

A Function for Array Addition

```
//This function computes the sum of two arrays a and b
//      and stores the sums in array c
void vecsum(float a[ ], float b[ ], float c[ ], int n){
   for(int i = 0; i < n; i++)
     c[i] = a[i] + b[i];
}
```

Note that in the function definition and prototype in Example 9.3, the length of the array c is given, whereas in Example 9.4 the lengths of the arrays are not given. Either construction is valid but giving only the array name is not. It is necessary to give at least the brackets [ ] so that the compiler knows that these are arrays. In calls of the function, however, only the array names are given, as was done in Example 9.3.

First Class Types   Although arrays may be used as function arguments, they may not be used as return values of a function. The fundamental types (int, float, and so on) may be used as not only function arguments but also return values and are called *first-class* types. Thus arrays are not first class in this sense. Values put into the array c of Example 9.4 are not returned as function values, but, rather, since the array c is called by reference, the values of the array c are placed directly in storage defined by main(). For example, suppose that the function of Example 9.4 has been called in main() by the statement

$$\text{vecsum(array1, array2, array3, 100);} \qquad (9.5)$$

where the first two arrays have been declared and assigned values in at least their first 100 positions and array3 has been declared to be of length at least 100. Then the address of array3 in the call (9.5) is used in place of c in the function, and the values of the sums are put directly into the corresponding positions of array3 in main().

Polynomial Evaluation

We end this section with another example of the use of arrays, as a means of holding the coefficients of a polynomial. Suppose that we wish to apply Newton's method, discussed in the previous chapter, to a polynomial equation $p(x) = 0$, where

$$p(x) = a_0 + a_1 x + \cdots + a_n x^n. \qquad (9.6)$$

Each iteration of Newton's method requires the evaluation of p and p' at the current iterate, where the derivative p' is

$$p'(x) = a_1 + 2a_2 x + 3a_3 x^2 + \cdots + na_n x^{n-1}. \qquad (9.7)$$

Consider the evaluation of $p(x)$ for some particular value x. A straightforward approach is to compute $x^2, x^3, \ldots, x^n$, and then evaluate (9.6). This requires n

Horner's Rule multiplications and n additions *plus* forming the powers of x. A more efficient algorithm is *Horner's rule*, also called the *nested form*, which for $n = 3$ is

$$p(x) = ((a_3x + a_2)x + a_1)x + a_0. \qquad (9.8)$$

Here, $a_3x + a_2$ is evaluated first, then multiplied by x, and so on. In general,

$$p(x) = (\cdots ((a_nx + a_{n-1})x + a_{n-2})x + \cdots + a_1)x + a_0. \qquad (9.9)$$

The evaluation of $p(x)$ by (9.9) requires n multiplications and n additions [check this for (9.8)], the same as (9.6) once the powers of x are known, but avoids the extra computation of the powers of x. It also usually has less rounding error. A function for the evaluation of $p(x)$ by (9.9) is given in Example 9.5. This function illustrates a backward `for` loop (Section 4.4) in which `i--` is the same as `i = i - 1`; this is just what is needed for the computation (9.9).

EXAMPLE 9.5
Polynomial Evaluation by Horner's Rule

```
float p(float x, float a[ ], int n){
  float value = a[n];
  for(int i = n-1; i >= 0; i--)
    value = (value * x) + a[i];
  return value;
}
```

Computing the Derivative The function of Example 9.5 may also be used to evaluate the derivative (9.6), but first we need to compute the coefficients of p'. Example 9.6 gives a function with the array a of coefficients of p as a parameter and produces an array der of coefficients of the derivative. This is a very simple example of a *symbolic differentiation* program in which the input represents a function and the output represents its derivative. (There are a number of programs, such as Mathematica, Macsyma, and Reduce, for symbolic differentiation.)

EXAMPLE 9.6
Obtaining the Derivative

```
void p_prime(float a[ ], float der[ ], int n){
  for(int i = 1; i <= n; i++)
    der[i - 1] = i * a[i];
}
```

In a program for Newton's method the function p of Example 9.5 may be called for evaluating the polynomial p, the function of Example 9.6 called for obtaining the array of coefficients of the derivative p', and then p again called by `p(x,der,n - 1)` to evaluate p'. This could all be combined into one function, of course, if desired (see Exercise 9.12).

MAIN POINTS OF CHAPTER 9

- Arrays are a way to define subscripted variables such as $a_0, \ldots, a_{n-1}$.
- The lengths of arrays must be specified when the arrays are declared.
- Arrays are usually used in conjunction with `for` loops or other repetition constructions.
- Arrays may be arguments of functions and are passed by reference, which means that only their address is passed to the function.
- An efficient way to evaluate polynomials is by Horner's rule.

EXERCISES

9.1   Run the programs of Examples 9.2 and 9.3 for various values of n, a, and b.

9.2   Modify the program of Example 2.1 so as to read n values of the radius r and store them in an array. Likewise store the corresponding computed values of v in an array.

9.3   Add `ofstream prt` and `prt` statements (Section 7.3) to the program of Example 9.2 so as to produce a file as output.

9.4   Write a function `main()` that will call the function of Example 9.4 for various values of n, a, and b.

9.5   Let $x = (x_1, x_2, x_3)$ and $y = (y_1, y_2, y_3)$ be two vectors starting from the origin in 3-space. The *dot product* (also called the *inner product*) of x and y is defined to be

$$(x, y) = x_1 y_1 + x_2 y_2 + x_3 y_3$$

Write a function to compute the dot product of two vectors x and y, where x and y are represented by arrays.

9.6   It is shown in analytic geometry that in terms of the dot product of Exercise 9.5 the cosine of the angle between two vectors u and v is

$$\cos \theta = \frac{(u, v)}{(u, u)^{1/2} (v, v)^{1/2}}$$

Write a function `main()` that will read in two vectors u and v, and compute $\cos \theta$ by calling the function of Exercise 9.5 three times. Then compute the angle θ by using the `acos()` function (arccosine) from `math.h`. Finally, print the vectors u and v and the angle θ in degrees. Test your program for a number of vectors u and v for which you can easily compute the exact answers.

9.7   Two vectors u and v are *orthogonal (perpendicular)* if their dot product is zero. But because of rounding error, two vectors that are orthogonal may yield a nonzero computed dot product. We will say that the vectors are *numerically* *orthogonal* if $|\cos\theta| \leq 10^{-6}$. Add to your program of Exercise 9.6 a test to determine if two given vectors u and v are numerically orthogonal, and, if so, print a suitable message to that effect.

9.8 Extend the programs of Exercises 9.5–9.7 to vectors u and v of length n. Declare the corresponding arrays in `main()` to be as large as the maximum value of n that you wish to consider.

9.9 Suppose that a is an array of floating point numbers satisfying `a[i+1] < a[i]` for $i = 0, \ldots, 99$. Write a program segment using a `while` loop to determine how many elements of the array a are ≥ 0.1.

9.10 The *mean* (average), $\bar{x}$, and *standard deviation*, σ, of n numbers $x_1, \ldots, x_n$ are defined by

$$\bar{x} = \frac{1}{n} \sum_{i=1}^{n} x_i, \quad \sigma = \left(\frac{1}{n} \sum_{i=1}^{n} (\bar{x} - x_i)^2 \right)^{1/2}$$

(Sometimes $n - 1$ is used in the denominator for σ.) Write a function that will accept n and the data $x_1, \ldots, x_n$ in the form of an array and compute $\bar{x}$. Then write a second function that will accept n, $\bar{x}$ and $x_1, \ldots, x_n$ and compute σ.

9.11 Let a and b be two one-dimensional arrays with ten elements each. Write a `for` loop that will copy the elements of a into b in reverse order, that is, `b[9] = a[0]`, `b[8] = a[1]`, and so on.

9.12 Combine the functions of Examples 9.5 and 9.6 into a single function that accepts the coefficients of a polynomial p and a value x as input and outputs the values of $p(x)$ and $p'(x)$. Write a main program that will call this function and test it for the polynomial

$$p(x) = 1.0 + 3.5x + 4.2x^2 + 32.0x^3$$

chapter
10

DIFFERENTIAL EQUATIONS

Many problems in science and engineering ultimately reduce to solving *differential equations*,[1] and solving such equations numerically is a very important problem in scientific computing.

10.1 THE INITIAL VALUE PROBLEM

In calculus, we learn how to solve simple differential equations such as

$$y'(t) = y(t) \qquad (10.1)$$

Here y is an unknown function of t and $y' = dy/dt$. The solution of (10.1) is

$$y(t) = ce^t \qquad (10.2)$$

for any constant c, as is verified by substituting (10.2) into (10.1) to produce the identity

$$ce^t = ce^t.$$

This example shows that the differential equation (10.1) by itself does not have a unique solution, since (10.2) is a solution for any c.

We may obtain a unique solution by requiring that the solution equal some specified value at some particular value of t. For example, if we require that the solution of (10.1) satisfy $y(0) = 3$, then the unique solution that achieves this is $y(t) = 3e^t$. In general, then, we will wish to solve a differential equation subject to the requirement that the solution at some point t be given. This prescribed value of the solution is usually called an *initial* condition, and the corresponding *initial value problem* is

---

1. If you have not yet studied differential equations, don't worry. You will only need to know what a derivative is.

$$\text{Solve } y'(t) = f(y(t)) \text{ subject to } y(a) = \alpha \qquad (10.3)$$

Initial Value Problem

In (10.3), a is a given value of t, α is a given number and f is a given function; for example, for (10.1), $f(y) = y$.

We were able to solve (10.1) explicitly because of its simple form, but this cannot be done for differential equations that arise in most applications. Rather, we must approximate the solution, and we next consider some simple methods to do this.

10.2 EULER'S METHOD

Suppose that $y(t)$ is the solution of the initial value problem (10.3). By definition of the derivative, we may approximate $y'(a)$ by

$$y'(a) \doteq \frac{y(a+h) - y(a)}{h} \qquad (10.4)$$

for sufficiently small h. Then, from (10.4), we have the approximation

$$y(a+h) \doteq y(a) + hy'(a).$$

Thus knowing $y(a) = \alpha$ by the initial condition, we may approximate y at $a + h$ if we know $y'(a)$. We don't know the solution y, but from the differential equation we have

$$y'(a) = f(y(a)).$$

Therefore,

$$y(a+h) \doteq y(a) + hf(y(a)) \qquad (10.5)$$

is a computable approximation to $y(a+h)$.

We may represent the approximation (10.5) geometrically. The equation of the tangent line to $y(t)$ at $t = a$ is

Geometry of Euler's Method

$$T(t) = y(a) + y'(a)(t - a).$$

Thus, at $t = a + h$, and again setting $y'(a) = f(y(a))$,

$$T(a+h) = y(a) + hf(y(a))$$

so that the approximation (10.5) is just the value of the tangent function at $a + h$. This is illustrated in Figure 10.1.

Once we have the approximation at $a + h$, we may repeat the process to obtain approximations for additional values of t. Let

$$t_0 = a, \quad t_1 = a + h, \quad t_2 = a + 2h, \ldots, t_n = a + nh \qquad (10.6)$$

Euler's Method

and $y_0 = y(a)$. Then, we may obtain approximations to the solution at the points (10.6) by

$$y_{i+1} = y_i + hf(y_i), \quad i = 0, \ldots, n \qquad (10.7)$$

This is *Euler's method*, the simplest method for approximating solutions of differential equations. A C++ program for Euler's method is given in Example 10.1. Note that arrays are used in this program to store the t_i and y_i values.

Figure 10.1 Approximation
Obtained from Tangent Line

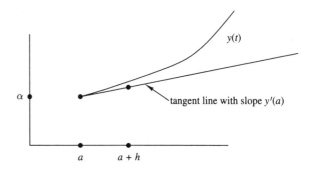

EXAMPLE 10.1
Euler's Method

```
///////////////////////////////////////////////////////////
// This program approximates a solution to a differential /
// equation y'=f(y) by Euler's method.  The function f is /
// assumed defined by a function subprogram.  The initial /
// point a, corresponding initial value y0, and the       /
// parameter h for Euler's method are input.  The values  /
// of t and the solution are stored in  arrays t and y    /
// dimensioned as 100 long.  Thus, n is restricted to no  /
// to no greater than 99 unless the array dimensions are  /
// changed.                                               /
///////////////////////////////////////////////////////////
#include <iostream.h>
#include <iomanip.h>
float f(float); //function prototype
int main(){
  float a,h,t[100],y[100];
  int i, n;
  cout << "values of a,n,h,y[0] are (n must be < 100)\n";
  cin >> a >> n >> h >> y[0];
  t[0] = a;
  for(i = 0; i < n ;i++){//this is the Euler loop
     y[i+1] = y[i] + h * f(y[i]);
     t[i+1] = t[i] + h;
  }
  //set headings for table and then print table
  cout << setw(2) << "t" << setw(6) << "y" << "\n";
  cout.setf(ios :: fixed);
  for(i = 0; i <= n ; i++)
  cout << setprecision(2) << setw(4) << t[i]
       << setprecision(4) << setw(8) << y[i] << endl;
  return 0;
}
  //this is the function f of the differential equation
float f(float y){
  return y;
}
```

TABLE 10.1
Euler Approximations for $h = 0.25$

t	y
0.00	1.0000
0.25	1.2500
0.50	1.5625
0.75	1.9531
1.00	2.4414

The function used in Example 10.1 is just $f(y) = y$ so that for this function the program will approximate solutions of $y'(t) = y(t)$. If the initial condition is $y_0 = 1$, and $a = 0$, then the exact solution of the differential equation is e^t. For $n = 4$ and $h = 0.25$, the output on the monitor is shown in Table 10.1. The headings for this table are set by a `cout` statement using the `setw` manipulator to obtain the correct spacing. Then the values of t and y are displayed, using the `setprecision` and `setw` manipulators.

Errors

The approximations to e^t in Table 10.1 are very poor; in particular, the solution for $t = 1$ should be $e = 2.71\ldots$. We next discuss the reasons for this large error.

There are two types of errors in Euler's method. The first is rounding error, which may occur in the evaluation of $f(y_i)$ as well as in the multiplications and additions in (10.7). The second, and usually more serious, error is discretization error, which results from replacing the derivative by the approximation (10.4). The smaller h is, the more accurate is this approximation and the smaller we would expect the discretization error to be. This is illustrated in Table 10.2 for the problem just discussed: $y'(t) = y(t)$ with $y(0) = 1$. Table 10.2 shows the approximate solution at $t = 1$ obtained by Euler's method with different values of h. The exact solution at $t = 1$ is $e = 2.718\ldots$, and the errors for the different values of h are given in the middle column.

Discretization Error

As the *step size h* decreases, the errors shown in Table 10.2 also decrease, as expected. The ratios of successive errors shown in the last column are decreasing

TABLE 10.2
Error in Euler's Method for $y' = y$ at $t = 1$

h	Computed value	Error	Error ratio
1	2.000	0.718	
1/2	2.250	0.468	0.65
1/4	2.441	0.277	0.59
1/8	2.566	0.152	0.55
1/16	2.638	0.080	0.53

more slowly, however. In fact, these error ratios will converge to $\frac{1}{2}$ as $h \to 0$. This is a consequence of the fact that the discretization error in Euler's method is proportional to h as $h \to 0$. Mathematically, we may write this as

$$|y_i - y(t_i)| \doteq ch, \text{ as } h \to 0, \tag{10.8}$$

Order of Method

where $y(t_i)$ is the exact solution of the differential equation at t_i, y_i is the Euler approximation at t_i and c is some constant. [The relation (10.8) is meant to hold at some *fixed* point $t_i = a + ih$, for example, the point $t = 1$ in Table 10.2. Thus, as $h \to 0, i \to \infty$ in such a way that t_i remains fixed.] Because of (10.8), Euler's method is called *first order* in h. A *higher-order* method will satisfy the relation

$$|y_i - y(t_i)| \doteq ch^p \tag{10.9}$$

for some integer $p \geq 2$, and p is called the *order* of the method. The higher the order, the more a decrease in h will decrease the discretization error. For example, if $p = 2$ and $h = 0.01$, then $h^2 = 0.0001$ so that the error is proportional to 0.0001, and not 0.01 as in Euler's method.

In practice, only high-order methods are used for solving differential equations, and Euler's method has been used in this section only for illustration. A second-order method is given in Exercise 10.2.

10.3 SYSTEMS OF EQUATIONS

We have considered so far only a single differential equation, but for most applications we must be able to solve systems of equations. Suppose that we have a system of *first-order* differential equations (only first derivatives are present), which we write in vector form as (vectors are in bold-face type)

$$\mathbf{y}'(t) = \mathbf{f}(\mathbf{y}(t)), \quad \mathbf{y}(0) = \mathbf{y}_0 \tag{10.10}$$

Here, there are n unknown functions $y_1(t), \ldots, y_n(t)$ with derivatives $y_1'(t), \ldots, y_n'(t)$ and n given functions $f_1(\mathbf{y}), \ldots, f_n(\mathbf{y})$ of the n variables $y_1, \ldots, y_n$ in the vector $\mathbf{y}$. Then the vectors in (10.10) are

$$\mathbf{y}(t) = \begin{bmatrix} y_1(t) \\ y_2(t) \\ \vdots \\ y_n(t) \end{bmatrix}, \mathbf{y}'(t) = \begin{bmatrix} y_1'(t) \\ y_2'(t) \\ \vdots \\ y_n'(t) \end{bmatrix}, \mathbf{f}(\mathbf{y}(t)) = \begin{bmatrix} f_1(\mathbf{y}(t)) \\ f_2(\mathbf{y}(t)) \\ \vdots \\ f_n(\mathbf{y}(t)) \end{bmatrix} \tag{10.11}$$

We may now apply Euler's method (10.7) to the system (10.10) as

$$\mathbf{y}_{k+1} = \mathbf{y}_k + h\mathbf{f}(\mathbf{y}_k), \quad k = 0, 1, \ldots \tag{10.12}$$

Thus, by use of vector notation, Euler's method has exactly the same form as for a single equation. Of course, for computation one still needs to deal with the components of the system. Thus, for $n = 2$, (10.12) would be written as

$$\left. \begin{array}{l} y_{1,k+1} = y_{1,k} + hf_1(y_{1,k}, y_{2,k}) \\[2mm] y_{2,k+1} = y_{2,k} + hf_2(y_{1,k}, y_{2,k}) \end{array} \right\} k = 0, 1, \ldots \tag{10.13}$$

A Predator–Prey Problem

As an example of a system of differential equations, we consider the interaction of two species that have a predator-prey relationship: The prey is able to find sufficient food but is killed by the predator whenever they encounter each other. An example of such species interactions is wolves eating rabbits. What we want to investigate is how the predator and prey populations vary with time.

Let $v(t)$ and $w(t)$ designate the number of prey and predators, respectively, at time t. To derive mathematical equations relating v and w we make several simplifying assumptions. First, we assume that the prey's birth rate, v_b, and natural death rate (exclusive of being killed by a predator), v_d, are constant with $v_b > v_d$. Thus the prey population, if left alone, increases at the rate $(v_b - v_d)v$. Second, we assume that the number of times that the predator kills the prey depends on the probability of the two coming together and is therefore proportional to vw. Combining these two assumptions, the prey population is governed by the differential equation

$$v'(t) = \alpha v(t) + \beta v(t)w(t) \tag{10.14a}$$

where $\alpha \equiv v_b - v_d > 0$ and $\beta < 0$.

In order to derive the predator equation, we assume that the number of predators would decrease by natural causes if the prey were removed, contributing a γw term. However, the number of predators increases as a result of encounters with prey, leading to

$$w'(t) = \gamma w(t) + \delta v(t)w(t) \tag{10.14b}$$

with $\gamma < 0$ and $\delta > 0$. In addition to these differential equations, we will have the initial conditions

$$v(0) = v_0, \quad w(0) = w_0$$

where v_0 and w_0 are the initial populations of the prey and the predators. Euler's method for the equations (10.14) is then obtained by taking $v = y_1$ and $w = y_2$ in (10.13) with the functions

$$f_1(\mathbf{y}) = \alpha y_1 + \beta y_1 y_2, \quad f_2(\mathbf{y}) = \gamma y_2 + \delta y_1 y_2 \tag{10.15}$$

Euler's Method for Predator–Prey Equations

Thus Euler's method for (10.14) is

$$\left. \begin{aligned} v_{k+1} &= v_k + h(\alpha v_k + \beta v_k v_k) \\ w_{k+1} &= w_k + h(\gamma w_k + \gamma v_k w_k) \end{aligned} \right\} k = 0, 1, \ldots \tag{10.16}$$

A C++ program to carry out (10.16) is given in Example 10.2. In this program we have denoted the number of prey, v, by `prey`, and the predators, w, by `pred`. The program sets α, β, γ and δ to constant values and reads in the value of h and the number, n, of Euler steps to take as well as the initial conditions for `prey` and `pred`. The `for` loop successively computes and displays the Euler approximations, rather than saving them in arrays as in Example 10.1. Note that since the current values of `prey` and `pred` are used in the temporary variable `temp`, they may be overwritten by the new values as they are computed. Note also that the first two `cout` statements in the `for` loop contain `flush` rather than `endl`. Thus there is no line spacing after these statements, and `time`, `pred`, and `prey` all print on one line. This could also have been accomplished by using the single `cout` statement

```
cout << " time = " << time << " pred = " << pred
     << " prey = " << prey << endl;
```

Exercise 10.3 asks you to obtain approximate solutions of the predator-prey equations by running the program of Example 10.2 for various values of *h*. Exercise 10.5 considers another system of differential equations, which model the trajectory of a projectile (for example, a cannon shell) that starts with some initial velocity and launch angle. This is a special case of the more general problem of computing rocket trajectories.

EXAMPLE 10.2
Euler's Method for Predator-Prey Equations (10.14)

```
///////////////////////////////////////////////////////////
//    This program obtains an approximate solution by      /
//    Euler's method to the equations (10.14) for given    /
//    values of alpha, beta, gamma and delta and given     /
//    initial values of the solutions prey and pred.       /
///////////////////////////////////////////////////////////
#include <iostream.h>
int main(){
  int n; //n is number of time steps to take
  float pred, prey, time = 0, h, temp;
  const float alpha = 0.25, beta = -0.01; //set parameters
  const float gamma = -1.0, delta = 0.01; //set parameters
  cout << "what are values of h and n?\n";
  cin >> h >> n;
  cout << "what are initial values of pred and prey?\n";
  cin >> pred >> prey;
  cout << "initial pred = " << pred << endl; //print initial
  cout << "  initial prey = " << prey << endl; //values
  for(int i = 1; i <= n; i++){
    temp = pred * prey; //current product
    prey = prey + h * (alpha * prey + beta * temp);
    pred = pred + h * (gamma * pred + delta * temp);
    time = time + h; //this is the current time
    cout << "time = " << time << flush;
    cout << "  pred = " << pred << flush;
    cout << "  prey = " << prey << endl;
  }
  return 0;
}
```

Modeling Error

Another very important type of error is not caused by errors in the computation but in the formulation of the problem. Many computational problems in science and engineering are *simulations* of some physical process. Examples of this are rocket or spacecraft trajectories, air flow over an aircraft, and blood flow in the human body. Such phenomena may be described by a *mathematical model*, which is usually one or more differential equations. The predator-prey equations (10.4) and the equations for the trajectory problem of Exercise 10.5 are examples of such mathematical models.

    The formulation of a mathematical model begins with a statement of the factors to be considered. In many physical problems, these factors concern the balance of forces and other conservation laws of physics. For example, in the formulation of a model of a trajectory the basic physical law is Newon's second law of motion, which requires that the forces acting on a body equal the rate of change of momentum of

Modeling Error   the body. This general law must then be specialized to the particular problem by enumerating and quantifying the forces that will be of importance. For example, the gravitational attraction of Jupiter will exert a force on a rocket in the earth's atmosphere, but its effect will be so minute compared to the earth's gravitational force that it can be neglected. Other forces may also be small compared to the dominant ones but their effects not so easily dismissed, and the construction of the model will invariably be a compromise between retaining all factors that could likely have a bearing on the validity of the results and keeping the mathematical model sufficiently simple that it is solvable using the tools at hand. Historically, only very simple models of most phenomena were considered since the solutions had to be obtained by hand, either analytically or numerically. As the power of computers and numerical methods has developed, increasingly complicated models have become tractable. However, all such models will still only be an approximation of reality.

EXERCISES

10.1 Run the program of Example 10.1 with values of h and n chosen so as to reproduce the values in Table 10.1. Keep halving h until the error in the approximation to e is less than 0.01. Do you need to modify the program to achieve this?

10.2 The *second-order Runge-Kutta method* for the initial value problem (10.3) is given by

$$y_{i+1} = y_i + \frac{h}{2}[f(y_i) + f(y_i + hf(y_i))], \quad i = 0, 1, \ldots$$

Modify the program of Example 10.1 to carry out this method. Run your program for the problem (10.1) with $y(0) = 1$ and compare your results with Euler's method. In particular, compare the accuracy of an approximation to e, as done in Table 10.1, for the same value of h in both programs.

10.3 Run the program of Example 10.2 for the initial values prey = 80, pred = 30, and for various values of h, for example, $h = 1, 0.5, 0.25$, and various values of n. The exact solution of the equations (10.14) is almost periodic for the given values of α, β, δ and γ and the initial conditions. How small do you need to take h before you begin to see this almost periodic behavior? If you have access to a graphics system, plot the solution in the pred, prey plane.

10.4 The second-order Runge-Kutta method of Exercise 10.2 may be written for the system (10.10) as

$$y_{i+1} = y_i + \frac{h}{2}[f(y_i) + f(y_i + hf(y_i))], \quad i = 0, 1, \ldots$$

Write this out explicitly for two equations and then write a C++ program to carry out this method for the predator-prey equations (10.14). For the same values as in Exercise 10.3, compare the accuracy with Euler's method. For what values of h do you begin to see the almost periodic behavior of the solution using the Runge-Kutta method?

10.5 The trajectory of a short-range projectile may be modeled by the following system of four differential equations:

$$x'(t) = v(t)\cos\theta(t), \qquad\qquad y'(t) = v(t)\sin\theta(t)$$
$$v'(t) = -\alpha[v(t)]^2 - g\sin\theta(t), \quad \theta'(t) = [-g\cos\theta(t)]/v(t).$$

Here, $x(t)$ and $y(t)$ are the coordinates of the projectile at time t, $v(t)$ is the velocity, $\theta(t)$ is the angle from the horizontal, g is the gravitational force, and α is a constant depending on air density, the coefficient of drag, mass, and cross-sectional area of the projectile. Euler's method for this system of differential equations is

$$\left.\begin{array}{l} x_{k+1} = x_k + hv_k\cos\theta_k \\[1.5em] y_{k+1} = y_k + hv_k\sin\theta_k \\[1.5em] v_{k+1} = v_k - h(\alpha v_k^2 + g\sin\theta_k) \\[1.5em] \theta_{k+1} = \theta_k - h\dfrac{g}{v_k}\cos\theta_k \end{array}\right\} k = 0, 1, \ldots$$

where $x_0 = y_0 = 0$ are the initial coordinates of the projectile, v_0 is the prescribed initial velocity, and θ_0 is the prescribed launch angle. Let $g = 9.81$ and $\alpha = 0.00215$, and write a program to carry out Euler's method.

a. If there is no drag, then $\alpha = 0$ and the differential equations have the exact solution

$$x(t) = (v_0\cos\theta_0)t$$
$$y(t) = (v_0\sin\theta_0)t - gt^2/2$$

Use this as a test case with the initial conditions $v_1 = 50$ m/sec and $\theta_0 = 0.6$ radians. Find a value of h that makes the computed solution accurate to approximately 2 decimal places.

b. For the original value of α, $v_0 = 50$ and the h you determined in a, find a value of θ_0 so that the projectile hits the ground at 125 meters. If you have access to a graphics system, plot this trajectory.

PART II

EXTENSIONS OF THE BASIC CONSTRUCTS

In Part I, we introduced many of the basic constructs and concepts of C++: variables and their types, assignment, the `if` statement for making decisions, the `for` and `while` loops for repetition, functions, arrays, and input/output statements. In Part II we return to some of these basic constructs for a more detailed and thorough treatment. We also introduce a number of new constructs and concepts.

In Chapter 11, we first discuss other fundamental data types and then treat operations in more detail. We also consider how to deal with non-numerical data (characters and strings of characters) and how to define other data types. In Chapter 12 we introduce two-dimensional arrays, which are used in scientific computing to represent matrices. A main scientific computing example in this chapter is the fundamental operation of matrix-vector multiplication. Chapter 13 covers the important scientific computing problem of solving systems of linear equations by Gaussian elimination. In Chapter 14 we return to functions and give a much more general treatment than was possible in Part I. Some of the topics here are reference parameters and return of references, recursive and inline functions, function over-loading, default parameters, and the construction and use of libraries of functions. Chapter 15 introduces the important idea of pointers, which are then used in Chapter 16 for dynamic memory allocation.

11

MORE ON DATA TYPES AND OPERATIONS

In this chapter, we will first return to the basic data types and arithmetic operations discussed in Part I and consider various extensions and restrictions on their use. We will then introduce a data type for characters and basic operations on characters and strings of characters. We also consider ways to define new data types.

11.1 OTHER FUNDAMENTAL DATA TYPES

In Chapter 1 we discussed how the size (also called the *range*) of integers depends on the number of bytes allocated. When we make the declaration

```
int i;
```

compilers on most PCs will allocate two bytes of memory for i (on Unix machines and 32-bit PCs, 4 bytes will usually be allocated, and sometimes 8 bytes) allowing integers in the range $-32,768$ to $32,767$ (see Tables 1.1 and 11.1). If this is not a large enough range, we can use the declaration

long *and unsigned* ints

```
long i;
```

which on most PCs will allocate four bytes of memory for i, allowing integers of magnitude approximately 10^9 (see Tables 1.1 and 11.1). There is also a short declaration to declare integers with less memory than int, but on some machines short also allocates 2 bytes of memory, the same as int. short is rarely used, and we will not consider it further.

Somewhat larger integers are possible if the integers are always positive. The declarations

```
unsigned int  i;
unsigned long j;
```

117

allocate the full number of bits to the magnitude, 16 in the first case and 32 in the second. Thus these types allow integers of twice the magnitude as the corresponding signed integers (see Tables 1.1 and 11.1).

For floating point numbers there are three possible levels of precision. The declaration

```
float x;
```

Higher Precision Floating Point

normally reserves four bytes of memory for x, allowing a precision of over 7 decimal figures and magnitudes in the range 10^{-38} to 10^{38}. The declaration

```
double x;
```

normally reserves eight bytes for x, which allows over 15 decimal digits of precision and a magnitude range of 10^{-308} to 10^{308}. These are called *double precision* numbers. Finally, the declaration

```
long double x;
```

normally reserves ten bytes for x. This allows a precision of roughly 19 decimal places and magnitudes of 10^{-4932} to 10^{4932}.

Example 11.1 gives a simple program to illustrate some of the above data types, and Table 11.1 summarizes the different types.

EXAMPLE 11.1
Defining Variables

```
#include <iostream.h>
#include <iomanip.h>
int main(){
    long i = 1234567; //define long integer i
    double x = 1.23456789876543; //define double precision x
    long double y = 1.234567898765432123; //define long double y
    cout << i << setprecision(20) << x << " " << y << endl;
    return 0;
}
```

Constants

Constants may also use different amounts of storage. For example, in the expression

```
i * 23L
```

the L specifies that 23 is to be represented in long format. Similarly, floating point constants are normally represented in double format, but in the expression

```
x + 2.34F
```

the F specifies that 2.34 is to be represented in `float` format.

Sometimes a program (or a programmer) may need to know exactly how much storage a given type requires on the system being used. This may be obtained by the `sizeof` operator. For example, the statements

The `sizeof` *operator*

TABLE 11.1
Summary of Numerical Data Types for PCs

Type	Bytes	Range	Precision (decimal)
int	2	$-32{,}768$ to $32{,}767$	—
long	4	-2×10^9 to 2×10^9	—
unsigned int	2	0 to 65,535	—
unsigned long	4	0 to 4×10^9	—
float	4	10^{-38} to 10^{38}	7
double	8	10^{-308} to 10^{308}	15
long double	10	10^{-4932} to 10^{4932}	19

```
int size_of_int = sizeof(int);

int size_of_double = sizeof(double);
```

will set `size_of_int` to the number of bytes that `int` variables require and `size_of_double` to the number of bytes for `double` variables. `sizeof` may also be used with variables themselves, as in

```
float x;
int i = sizeof(x); //same as sizeof(float)
```

Synonyms Using `typedef`

A sometimes useful practice is to define synonyms for the above types, which is done by means of the `typedef` statement. For example, the statement

```
typedef float Real;
```

defines `Real` to be a synonym for `float`. Thus the subsequent statement

```
Real d;
```

declares `d` to be a variable of type `float`. The advantage of using `typedef` is that uniform changes to type may easily be made. For example, suppose that you wished to run a program with all `float` variables changed to `double`. Normally, you would have to find all declarations of `float` variables in the program and change them, a time-consuming and error-prone task if the program is large. However, if `typedef` had been used to define a synonym for `float`, as is `Real` in the example above, then only the single `typedef` statement would need to be changed:

```
typedef double Real;
```

Then all occurrences of `Real` in the program would automatically be made `double`.

11.2 OPERATIONS

In this section we collect a number of additional facts about arithmetic operations including overflow, mixing of different types and type casting, precedence, compound operations, and bitwise operations.

Overflow

Floating Point Overflow

If we multiply two numbers of type `float`, we will overflow if the product is greater than the maximum allowed. For example, if the maximum is about 10^{38} and if we have the statements

$$\text{float a = 1.0e+20, b = 3.14e+20;} \qquad (11.1a)$$

$$\text{a = a * b;} \qquad (11.1b)$$

we will obtain a floating point overflow in computing `a * b`. If we anticipate this possibility, we can declare the variables to be `double`. Thus, if we changed the `float` in (11.1a) to `double`, then a and b could contain magnitudes up to about 10^{308}, and there would be no overflow. Of course, we could overflow using `double` if the magnitudes of a and b were sufficiently large, for example, 10^{160}.

Integer Overflow

On most systems (but not all), when overflow occurs with floating point operations, this is a run-time error and the program stops. But the same may *not* happen with integer arithmetic. For example, consider the statements

```
int i = 32000;
i = i + 1000;
```

This will cause an overflow if the maximum integer allowed by the `int` declaration is 32,767 (see Table 11.1). However, there may be no run-time error indication, and an erroneous result will be stored for i. (The result stored will be $-32,536$ because of the way integer arithmetic is done.)

Mixing Types

Suppose that instead of (11.1a) we had the statements

```
double a = 1.0e+20;
float b = 3.14e+20;
a = a * b;
```

Then the multiplication a * b *mixes* the `float` type b with the `double` type a. This is permissible: b will be temporarily converted to `double`, and the result of the multiplication is stored in a as `double`. The same principle applies whenever we mix different types in an arithmetic expression. For example, consider the statements

```
int  i = 23;

float  x = 14.2,  y;                                    (11.2)

y = i + x;
```

Here, `i` is temporarily converted to `float` for the addition, and the result stored as type `float` in `y`. In general, types will be converted according to the hierarchy (left to right)

Numerical Hierarchy

$$\texttt{long double, double, float, long, int} \qquad (11.3)$$

Thus, if two variables of different types are mixed, the value of the one to the right in the hierarchy (11.3) is temporarily converted to the type on the left for the operation. Such a temporary conversion is only for the operation being done and does not affect the variable in storage. Thus in (11.2) `i` remains an integer in storage.

Conversion of Function Parameters

Another context for automatic type conversion involves functions. Suppose that a function has the prototype

```
float f(float);
```

and that it is called by the statement

```
i = f(j);
```

where `j` is of type `int`. Then there is a mismatch between the type of the argument as specified in the function interface and the type of the actual parameter. In this case, the value of `j` passed to the function is converted to type `float` as a local variable in this function, although `j` remains an integer in the calling program.

We note that all the mathematical functions in the library `math.h` return a function value of type `double` and usually expect an argument of type `double`. If an argument of type `float` is given, conversion to type `double` will be automatic. Also, if assignment of the function value is made to a variable of type `float`, the conversion from `double` to `float` is automatic.

Casting

Although conversions such as `i` to `float` in (11.2) are done automatically, they may also be done explicitly. For example, (11.2) could be written as

```
y = float(i) + x;                                       (11.4)
```

This is called *casting* or *typecasting*. Although (11.4) has exactly the same effect as (11.2), the expression

```
i + int(x)                                              (11.5)
```

x Is Truncated

does a conversion that cannot be done automatically according to the hierarchy (11.3). The effect of (11.5) is to truncate (*not* round) `x` temporarily to an integer that is then added to `i`. For example, if `i = 21` and `x = 4.63` before (11.5) is executed, then the sum will be $21 + 4 = 25$.

Quotients of Integers

Recall that if `i` and `j` are integers, the division `i/j` produces only the integer part of the quotient. If your intent is to obtain the correct quotient, you can cast one of the operands as in:

```
float y = 1/float(2); //stores .5 in y
```

Or you can write one of the operands as a floating point integer

```
float y = 1/2.0; //stores .5 in y
```

Here, 2.0 is interpreted as a floating point number so that the division is done in floating point arithmetic.

Assignment Conversion

Another type of conversion involves assignment. If i is type int and y is type float, the assignment

```
y = i; //stores float(i)
```

will store i in y as type float. Similarly, the assignment

```
i = y; //stores int(y)
```

will store the integer part of y in i.

Run-Time Errors

Conversions that cannot be successfully made may cause run-time errors. For example, the float number with magnitude 12345678.1 cannot be converted to type int because it is too large. Similarly, conversion of a double number with exponent greater that 10^{38} cannot be made to float. If conversions of this kind are attempted, you will receive an error message like "Floating Point: Invalid."

Precedence

Binary and Unary Operators

The operators / and * are called *binary operators* since two operands are always required, as in a/b and c * d. + and −, on the other hand, may be either binary, as in a + b and c - d, or *unary*, in which there is a single operand, as in +a or -b. The operators /, *, +, - are called *left associative* since in expressions such as a/b * c or a + b - c containing operators of the same precedence level, b *associates* with the operator to the left, which is carried out first.

Multiple Assignment

Another type of precedence involves *multiple assignment*. The statement

```
a = b = c;
```

is valid and stores the current value of c into both a and b. In the statement

```
a = b = c + 4;
```

the computation of c + 4 is given precedence and is computed first; then it is stored in both a and b.

Incrementing

We have used the notation k++ to mean k = k + 1, but we can also use k++ in expressions. Consider the expression

$$i = j * k++; \tag{11.6}$$

for integers i, j, and k. This statement will form the product of j and k and then increment k by 1. Thus (11.6) is equivalent to the two statements

$$i = j * k;$$
$$k = k + 1;$$

Pre- and Postincrements

On the other hand, in the statement

$$i = j * ++k; \tag{11.7}$$

k is incremented *before* the multiplication; thus (11.7) is equivalent to the two statements

$$k = k + 1;$$
$$i = j * k;$$

k++ is called a *postincrement* since k is incremented after it is used and ++k is called a *preincrement* since k is incremented before it is used. The *decrement* operators k-- and --k behave in the analogous way except that now k is replaced by k - 1. Note that we cannot use the increment or decrement operations on expressions; for example, (i + 2)++ is not legal.

Compound Assignment Operations

i++ is one of many "shorthand" expressions allowed in C++. Another type of shorthand expression is called a *compound assignment operator* and is illustrated by

$$i \mathrel{+}= j \text{ is same as } i = i + j \tag{11.8a}$$

The same construction may be used with *, /, %, and - :

$$i \mathrel{*}= j \text{ is same as } i = i * j \tag{11.8b}$$
$$i \mathrel{/}= j \text{ is same as } i = i/j \tag{11.8c}$$
$$i \% = j \text{ is same as } i = i \% j \tag{11.8d}$$
$$i \mathrel{-}= j \text{ is same as } i = i - j \tag{11.8e}$$

The variables in (11.8) may be of any type: int, float, etc. Note that in (11.8a) there can be no space in +=, and similarly in the other constructions of (11.8).

The shorthand expressions (11.8) are widely used by C++ programmers, partly because they may lead to more efficient programs. But you may prefer to ignore them.

Bitwise Operations†

C++ allows operations on the individual bits of integers. The operations that are allowed are shown below, illustrated by two four-bit binary numbers:

---

†This subsection may be omitted without loss of continuity.

```
~ bitwise negation    ~0110           =  1001
& bitwise And         0110 & 1100     =  0100
| bitwise Or          0110 | 1100     =  1110
^ exclusive Or        0110 ^ 1100     =  1010
```

In the bitwise exclusive Or the result is 1 if either but not both bits are 1; otherwise the result is 0.

These operations may also be used on variables of type int, not just constants, as illustrated by

$$i = {\sim}j; \; i = j \; \& \; k; \; i = j \; | \; k;$$

It is also possible to shift bits, either right or left. For example, if i has the binary representation 101101, the statement

$$j = i \; << \; 2;$$

will shift the bits of i two places left so that j will be 10110100, where zeros have been entered from the right. Similarly

$$j = i \; >> \; 2;$$

would shift the bits of i two places right so j would be 1011. Note that any bits shifted beyond the length of the word are lost.

Since bitwise operations are used primarily in systems software, such as operating systems, we will not pursue them any further.

11.3 CHARACTERS AND STRINGS

In addition to the types int, float, and so on for numerical data, C++ also has a data type for characters. The statement

$$\text{char} \quad \text{c}; \tag{11.9}$$

declares c to be a *character variable* and c may be assigned characters as values, as in

$$\text{c = 'a'}; \tag{11.10}$$

Here, the ' ' surrounding a indicate that it is the letter a, and not a variable. Recall from Chapter 1 that a character is normally stored in one byte of memory, using the ASCII codes of Appendix 1. Thus the declaration (11.9) reserves one byte of memory [hence, sizeof(char) = 1] for c, and (11.10) assigns the ASCII code for a to this memory location. These ASCII codes are positive binary integers, and C++ allows characters to be treated as 8-bit integers. In particular, an arithmetic statement like

$$\text{c = 'a' + 5};$$

is permissible and stores the binary code for the character `'f'` in c. (See Appendix 1 for the ASCII codes for `'a'` and `'f'`.)

Strings

Usually we will be interested in working with *strings* of characters, rather than single characters. For example, the sentence

$$"I\ am\ a\ baker" \qquad (11.11)$$

is a string of 12 characters, including the three blanks. Many programming languages allow string variables, but C++ does not. Rather, strings are held in character arrays. The statement

$$char\quad c\_str[4]; \qquad (11.12)$$

Strings as Arrays
declares c_str to be an array of 4 characters. We may initialize it in the same way as a numerical array:

$$char\ c\_str[4] = \{'a',\ 'b',\ 'c'\}; \qquad (11.13)$$

This statement declares c_str to be a character array of four elements and sets the first three characters to a, b, and c. This initialization may also be achieved by

$$char\ c\_str[4] = "abc"; \qquad (11.14)$$

Note that here we use double quotes, rather than the single quotes of (11.13), and no braces are necessary. As another example, the statement

$$char\ baker[13] = "I\ am\ a\ baker"; \qquad (11.15)$$

declares baker to be an array of 13 characters and initializes the first 12 characters to the indicated sentence, including the blanks. When we initialize the array, we do not need to specify the length if we wish it to be the same length as the initializing string; thus we could write (11.15) as

$$char\ baker[\ ] = "I\ am\ a\ baker";$$

Each string represented as an array must terminate with the *null character*, which is the character whose ASCII code is zero. In most cases, as in (11.15), this null character will be automatically attached. In those cases in which it is entered *Beware the Termination Character* explicitly, it is done as in

$$char\ c\_str[4] = \{'a',\ 'b',\ 'c',\ 0\};$$

Using `'0'` would not work here since the ASCII code for the character `'0'` is 30, and not 0. `'\d'` may also be used. Note that room for this termination character must *Leave Room for 0* be allowed in the declaration of the array. Thus in (11.12) and (11.15) we declared the array to be one longer than the character strings to be entered in them. If a global character array is not explicitly initialized, as in (11.13) or (11.14), the compiler will initialize all characters to be the null character. Local arrays are not automatically initialized.

Using String Arrays

Once a character array has been declared, we may use it as we would use any other array. For example, if c_str and baker have been declared by (11.12) and (11.15), the statement

```
c_str[0] = baker [3];
```

puts m, the fourth character of the string in the array baker, into the first position of c_str. Similarly,

```
c_str[1] = 'b';
```

puts the character b into the second position of c_str. In general, we must use subscripts to access the elements of an array so that statements like

```
c_str = 'abc'; //not legal

c_str = "abc"; //not legal
```
(11.16)

are not legal. Also, as with other arrays, a statement such as

```
c_str = baker; //not legal
```

is not legal, even if c_str is defined to have the same length as baker.

Reading and Printing Strings

Reading a Character

Character variables may be read and displayed by cin and cout, just as numbers. Consider the statements

```
char c;
cout << "what is c?\n";
cin >> c;
cout << c << endl;
```

Reading and Displaying a String

Here, the cin statement will read a character from the keyboard and assign it to the character variable c. Then the cout statement will print the character in c. We may also read and print strings in much the same way, as illustrated in Example 11.2.

EXAMPLE 11.2
Reading a String

```
char c_str[4];
cout << "what is c_str?\n";
cin >> c_str; //read whole string
cout << c_str << endl;
```

The cin statement in Example 11.2 will now expect up to 3 characters from the keyboard (the fourth character being reserved for the termination character). If you type in abc, these 3 characters will be assigned to the array c_str and then be displayed on the monitor by the cout operation. If you type in only 2 characters, say ab, these will be assigned to the first two positions of the array, and the cout statement will print only these two characters.

Example 11.2 illustrates that with characters just an array name, c_str in this case, may be used in cin or cout statements. But the same cannot be done with numerical arrays. For example, the statements

```
float arr[2] = {6.5, 4.8};
cout << arr; //does not display a
```

will compile, but the cout statement does not give the desired result of displaying the two elements of a. Moreover, the statements

```
float arr[2];
cin >> arr; //error
```

give a syntax error.

Next, suppose that you wished to enter a b, with a blank between a and b. Unfortunately, this will not work since the cin operator skips leading blanks but after reading a character treats a blank as a termination character and will end the read with only a entered into c_str[0]. We can solve this problem by replacing the cin operation by

Problem with Blanks

$$\text{cin.get(c\_str,4);} \qquad (11.17)$$

Now, any three characters, including blanks, may be entered from the keyboard and stored in the array c_str. (Fewer characters may be entered by terminating with a return.) Use of get(), which is a function included in the iostream.h file, has another advantage. Suppose that with the program in Example 11.2, you type in the five characters doggy. The array c_str has been declared as having only 4 elements, but there is no protection mechanism to stop storing characters when the array size is reached. Thus the fifth character will be stored in the location following c_str[3], possibly overwriting some of your other data. The use of (11.17) would prevent this, since a maximum of 4 characters (including the termination character) can now be read.

The overwriting problem could be prevented in another way, by using the setw manipulator previously discussed in Section 7.2 in conjunction with the cout operation. If the cin statement in Example 11.2 is replaced by

$$\text{cin >> setw(3) >> c\_str;} \qquad (11.18)$$

then a maximum of 3 characters will be read. As always, the manipulator file iomanip.h must be used with setw.

The Carriage Return Problem

Even though a blank will not terminate a cin.get() statement, a carriage return will. If you are typing in a long sentence, or several sentences, it may well be necessary to use several lines, and you do not want your input prematurely terminated by a carriage return. We can resolve this problem by using the following more general form of cin.get():

```
cin.get(c_str,300,'?')
```

where ? is a termination character. Now characters will be read until either 300 characters have been entered, or the character ? is entered; a carriage return no longer terminates the input. ? may be replaced by any character you choose, one that would not occur in your input. Note that each carriage return is now entered as a character in the string, so that space must be allowed for these. Also, if the string is printed by cout, these carriage returns will cause the same line spacing as in the input string.

Problem with There is a problem with cin.get() when it is to be used for successive reads.
cin.get For example, suppose we have the loop

```
for(int i = 0; i < 5; i++){
  cin.get(c_str, 3);
  cout << c_str << endl;
}
```

The second time through this loop, cin.get() will not stop to accept any input. The reason is that the carriage return used to terminate the first string is left in the "input stream" and is the first character read during the second use of cin.get(). This terminates the read before any other characters can be typed. One way to circumvent this problem is to use the function cin.getline() in place of cin.get(). Thus, in the loop

```
for(i = 0; i < 5; i++){
  cin.getline(c_str,3);
  cout << c_str << endl;
}
```

there is now no problem with successive reads. As opposed to cin.get(), cin.getline() extracts the carriage return from the input string.

get() and getline() may also be used with file input. If, for example, read has been defined by the statement (see Section 7.3)

```
ifstream read("data");
```

then the statements

```
read.get(c_str, 3);
```

and

```
read.getline(c_str, 3);
```

behave in exactly the same way as cin.get() and cin.getline() except that characters are now being read from a file rather than the keyboard.

We summarize the properties of cin, cin.get(), and cin.getline() that we have discussed for reading character strings:

- cin is terminated by a blank as well as a carriage return.
- cin.get(arrayname, n) puts up to n characters, including blanks, into arrayname. Terminated by return.

- `cin.get(arrayname, n, character)` terminated by character instead of carriage return.
- `cin.getline(arrayname, n)` same as `cin.get` except carriage return is not left in stream.

Characters (but not strings) may also be used with the `switch` construct (Section 3.3), as in:

```
switch(ch) {
  case 'A': cout << "A";
    break;
  case 'B': cout << "B";
    break;
  default: cout << "no match";
}
```

Characters as Switch Variables

In this example, it is assumed that `ch` is a `char` variable. The switch statement attempts a match with the characters 'A' and 'B'; if no match is found, the default statement is executed.

11.4 USER-DEFINED DATA TYPES

We have so far discussed a number of data types: `float`, `int`, `char`, and so on; these data types are called *fundamental*, *primitive*, or *built-in* types and are part of the language. C++ also allows additional *user-defined* data types by means of *enumerated types, structures*, and *classes*. This section discusses the first two of these and classes will be covered in Part III.

Enumerated Data Types

The latest C++ standard has a data type, `bool`, that takes on only two possible values, `true` or `false`. Thus the declaration

$$bool\ logical = true;$$

defines a `bool` variable named `logical` and initializes it to the value `true`. A `bool` variable may be used in assignment statements such as

$$logical = a > 0;$$

Here `logical` will be assigned the value `true` if `a > 0` and `false` otherwise.

By means of *enumeration*, it is possible to define other types that take on only a limited number of values. For example, the statement

$$enum\ shapes\{circle,\ square,\ rectangle,\ polygon\}; \quad (11.19)$$

defines a type `shapes` that can take on the four indicated values.

The values of the `enum` list are represented by the integers 0, 1, 2, ... from left to right. Thus the four values in (11.19) are represented by 0, 1, 2, and 3, although these integers may be given other values by, for example,

```
enum shapes{circle = 3, square = 5, rectangle = 8, polygon = 9};
```

Since the values are integers, it is possible to do integer arithmetic with them, although in most cases this will not make any sense. However, suppose that the enumerated list is

```
enum days{sunday, monday, tuesday, wednesday,
          thursday, friday, saturday};
```

Then after the declaration

```
days day1 = monday, day2 = wednesday;
```

the expression

```
day2 - day1
```

would be evaluated as `3 - 1 = 2`, giving the number of days between day1 and day2.

Structures

A simple example of a structure is given in Example 11.3. The structure is called `car`, and this name appears in the first line of the definition. `weight`, `length`, and `id_number` are the *members* of the structure. Note that this structure is somewhat like an array, in that there are three data members. However, two members are of type `float` and the other of type `long`, whereas in an array all elements must be the same type. The list of member names is enclosed in braces and a semicolon *must* follow the closing brace.

EXAMPLE 11.3
A Structure

```
struct car{
  float weight;
  float length;
  long id_number;
};
```

Declaring a Structure Variable and Setting Member Values

After a structure is defined, as in Example 11.3, variables of that structure type may be declared by a statement such as

$$car \quad dodge; \tag{11.20}$$

This statement declares `dodge` to be of type `car` and reserves storage for its three members. At this point the members have not been assigned values, but we may do this by assignment statements such as

$$\text{dodge.weight = 3020.;} \qquad (11.21a)$$

$$\text{dodge.length = 196.;} \qquad (11.21b)$$

$$\text{dodge.id\_number = 54867645;} \qquad (11.21c)$$

The *dot operator* (*member access operator*) in (11.21) plays much the same role as brackets in an array; for example, if `arr` is an array, `arr[2]` is the third element, but in structures individual members are addressed by their name, as in (11.21), as opposed to their position in the structure. However, just as with arrays, structures may be initialized by a statement of the form

```
car dodge = {3020., 196., 54867645};
```

This statement would combine the declaration (11.20) with the initializations of (11.21).

The definition of a structure may be combined with the declaration of a variable of that type. If we change Example 11.3 to

A Restrictive Definition

```
struct{
  float weight;
  float length;
  long  id_number;
} dodge;
```

then (11.20) is no longer needed. However, we do not recommend this construct, although you may see it used occasionally. It restricts us to a single variable, `dodge`, of this structure type, whereas the definition of Example 11.3 allows us to declare as many variables of this type as we like. For example

```
car ford, chevy;
```

will define two more variables of type `car`.

If `ford` and `dodge` are two structures of type `car`, we may treat their members as individual variables by means of the dot operator. Thus, if (11.21a) has assigned a value to the first member of `dodge`, the statement

Working with Members

```
ford.weight = dodge.weight;
```

will assign the same value to the first member of `ford`. However, we can assign all members at once by the statement

```
ford = dodge;
```

The individual members may also be used in arithmetic expressions such as

```
dodge.weight + ford.weight
```

or in input/output statements like

```
cin >> dodge.weight;
cout << dodge.weight << endl;
```

A Structure May Be a Structure Member

In general, we may use the individual members in the same way as any other variables.

Members of a structure may themselves be structures. For example, suppose that we define another structure:

```
struct weight_parts{
    float trans_weight;
    float engine_weight;
    float rest_weight;
};
```

Then we replace the declaration of `weight` in Example 11.3, declaring it to be of type `weight_parts`. We illustrate this by the program shown in Example 11.4.

EXAMPLE 11.4
A Program Using Structures

```
#include <iostream.h>
 struct weight_parts{ //define a structure
    float trans_weight;
    float engine_weight;
    float rest_weight;
 };
 struct car{ //define another structure
    weight_parts weight; //this member is a structure
    float length;
    int id_number;
 };
 int main(){
    weight_parts ford_weight = {400.,800.,1600.};
    car ford;
    ford.weight = ford_weight;
    cout << ford.weight.trans_weight << endl;
    cout << ford.weight.engine_weight << endl;
    cout << ford.weight.rest_weight << endl;
    return 0;
 }
```

In Example 11.4, we define the structure `car` whose first member is a structure of type `weight_parts`. Note that these structure definitions appear before `main()`. They could be inside `main()` but must come before a variable of that structure type is declared. We declare `ford_weight` to be a structure of type `weight_parts` within `main()` and initialize it. After declaring `ford` to be a structure of type `car`, we set its `weight` member equal to the structure `ford_weight`. Finally, we print the three components of `ford_weight`. Note that here we use the dot operator twice; thus, for example,

```
ford.weight.trans_weight
```

gives the `trans_weight` member of the structure `ford.weight`.

We may also define arrays of structures, as shown in Example 11.5. In this example, the weight member of each structure is read by using `.weight` with the structure `cars[i]`, which is the `(i + 1)` structure in the array. Then the weight is displayed.

EXAMPLE 11.5
An Array of Structures

```
#include <iostream.h>
struct car{ //define the structure
   float weight;
   float length;
   int id_number;
};
const int max = 3; //array size
int main(){
   car cars[max]; //declare array of structures
   for(int i = 0; i < max; i++){
     cin >> cars[i].weight;
     cout << cars[i].weight;
   }
   return 0;
}
```

We note that structures are first class types and enjoy all the properties of fundamental types; besides being used in arrays, they may be return values or parameters of functions, and so on. In Part III, we will see that classes are a very important extension of structures.

MAIN POINTS OF CHAPTER 11

- The data types `long`, `unsigned int`, and `unsigned long` allow for larger integers than `int`. Likewise, `double` and `long double` allow floating point numbers with more precision and larger magnitudes than `float`.
- Overflow will usually cause run-time errors and program termination with the `float`, `double`, and `long double` types. Overflow may also occur with the integer types and produce erroneous results but not program termination.
- Types may be mixed in arithmetic expressions and automatic temporary conversion to the highest level type will be done. Explicit conversion may also be done by casting.
- Many operations may be written in a shorthand form, for example, `a += b` for `a = a + b`.
- `char` is the data type for characters. Strings of characters may be defined as arrays of characters.

- Characters may be read and printed by the `cin` and `cout` operations. They may also be read by the `cin.get()` and `cin.getline()` functions.
- Structures are user-defined data types that may consist of more than one component, not all the same type.
- Data types may also be defined by enumeration.

11.1   Suppose that you wish an integer variable `i` to be able to take on values in the range $-100,000$ to $100,000$. How would you declare `i`? If `i` were to take on values in the range 0 to 50,000, how could you declare `i` to minimize the storage it requires?

11.2   Suppose you wish a floating point variable `r` to have at least 10 decimal digits of precision. How would you declare `r`? What if you wanted it to have at least 17 decimal digits of precision?

11.3   Run the following programs and comment on their different behaviors. [Part b assumes integers are stored in two bytes, not four.]

    a.
```
int main(){
    float x = 3.14e+20;
    x = x * x;
    cout << x << endl;
    return 0;
}
```
    b.
```
int main(){
    int i = 30000;
    i = i * i;
    cout << i << endl;
    return 0;
}
```

11.4   Give the results of the following operations if `i` = 2 and `j` = 3 are integers, and `x` = 1.5 is floating point.

    a. `x = i + j;`
    b. `i = j + x;`
    c. `i = j + int(x);`

11.5   Explain the meaning of each of the following statements, where `i`, `j`, and `k` are integer variables.

    a. `i += j;`
    b. `i = j + k++;`
    c. `i *= j;`

11.6   Let a character string be defined by

```
char str[20] = "CS101_is_great";
```

What do the following `cout` statements display?

a.   `cout << str[3];`

b.   `for(int i = 0; i <= 15; i = i + 2)`
        `cout << str[i];`

11.7   Write a program that will read in a string of up to 20 characters from the keyboard and also read a second string of one character, called the test character. Then count the number of occurrences of the test character in the first string. Output the two strings and the count.

11.8   Define a structure, `plane`, whose members are the weights of the wings, the fuselage and the remaining weight. Then write statements that assign values to these members.

11.9   Define a structure for complex numbers, where the members of the structure are the real and imaginary parts.

11.10   Define a structure for rectangles that consists of two members: height and width. Then declare an array of such structures and write statements to input values by `cin` and display values by `cout`.

11.11   Define an enumerated data type by

```
enum car{weight, length, id_number};
```

Discuss how this data type differs from the structure `car` of Example 11.3.

chapter
12

LOTS MORE VALUES: TWO-DIMENSIONAL ARRAYS

Much of scientific and engineering computation involves vectors and matrices. In this chapter, we introduce two-dimensional arrays as the natural way to represent matrices and give as a main example some programs and functions to carry out matrix-vector multiplication. We also use two-dimensional arrays for character strings.

12.1 TWO-DIMENSIONAL ARRAYS

A vector is defined in C++ by the arrays we have already discussed. For example, the vector with real elements

$$\mathbf{x} = \begin{bmatrix} x_1 \\ \vdots \\ x_{10} \end{bmatrix} \tag{12.1}$$

would be represented by

```
float  x[10];                        (12.2)
```

Matrices   An $m \times n$ *matrix* A is of the form

$$A = \begin{bmatrix} a_{11} & \cdots & a_{1n} \\ \vdots & & \\ a_{m1} & \cdots & a_{mn} \end{bmatrix} \tag{12.3}$$

with m rows and n columns. Two examples are

$$A = \begin{bmatrix} 2 & 1 & 2 \\ 3 & 2 & 4 \end{bmatrix} \quad A = \begin{bmatrix} 1 & 1 & 2 \\ 4 & 2 & 1 \\ 6 & 4 & 2 \end{bmatrix} \tag{12.4}$$

136

Two-Dimensional Arrays

with $m = 2$ and $n = 3$ in the first case, and $m = n = 3$ in the second. Matrices may be represented in C++ by *two-dimensional arrays* defined by declarations such as

$$\text{float A[2][3];} \tag{12.5}$$

This declares A to be an array with 2 rows and 3 columns so that it represents a 2×3 matrix. Integer arrays may be defined in a similar way: The statement

```
int int_array[4][2];
```

declares int_array to be a 4×2 array of integers. And two-dimensional arrays of type double, long, char, and so on, may also be defined. Arrays with only one index, such as x in (12.2), are called *one-dimensional* arrays.

In Section 11.1 we discussed the sizeof function, and it can be used for arrays also. For example, if x is defined by (12.1), the value of sizeof(x) would be $4 \times 10 = 40$, assuming that float requires 4 bytes of storage. Similarly, if A is defined by (12.5), the value of sizeof(A) would be $2 \times 3 \times 4 = 24$.

Array Indices

There is a certain problem working with vectors and matrices in C++. The usual mathematical notation for vectors and matrices, illustrated in (12.1) and (12.3), is that the components start with the subscript 1. This conflicts with the C++ convention that array indices start with the index 0. In order to retain the mathematical convention, we could ignore the 0 index in arrays when working with vectors and matrices. This means, however, that we must declare the arrays to be one longer in each dimension. Thus, for example, we would replace (12.2) by

```
float x[11];
```

The array **x** is now 11 long, and the element x[i] will correspond to x_i in the vector **x** of (12.1). We would simply ignore the element x[0] in the array.

Another approach would be to define a function that would shift the array index. For example, the function

```
float element(float x[ ], int i){
    return x[i-1];
}
```

Shifting Indices

has as arguments an array x and integer i and will return the array element x[i-1]. Thus, the statements

```
float x[10], y;
y = element(x, 4);
```

would set y equal to the fourth element of **x**, x_4, consistent with the mathematical notation.

Both of these approaches are rather clumsy and unsatisfactory, and for now we will simply use the indices starting with 0, which means that x[i] will be the element x_{i+1}. We will see later how we can use pointers and classes to keep the usual mathematical indexing.

Higher-Dimensional Arrays

We can also declare arrays of more than two dimensions. For example,

```
float big[10][10][10];
```

defines a three-dimensional array of $10 \times 10 \times 10 = 1000$ elements. Arrays of still higher dimensions are possible, although more than three dimensions are rarely used.

Initialization of Arrays

Two- and higher-dimensional arrays may be initialized in much the same way as one-dimensional arrays. For example, the statement

$$\texttt{float A[2][3] = \{\{2, 1, 2\},}$$
$$\texttt{\{3, 2, 4\}\};} \qquad (12.6)$$

declares and initializes a 2×3 array. The values of each row of the array are given in order; in (12.6) we have listed the second row on a second line, to give the appearance of a matrix, but this is not necessary. In fact, we may write (12.6) as

```
float A[2][3] = {2, 1, 2, 3, 2, 4};
```

with no additional braces. As with one-dimensional arrays, unless specifically initialized otherwise, global two- and higher-dimensional integer and floating point arrays are automatically initialized to 0.

Storage of Arrays

For many purposes, it is important to know how arrays are actually stored in memory. For one-dimensional arrays there is no problem: for the array x declared by

```
float x[10], A[3][3];
```

10 consecutive memory locations are reserved for x. For two-dimensional arrays, however, there are different possibilities. In C++, storage is *row-major*; that is, the first row is stored, followed by the second row, and so on, as illustrated in Figure 12.1. Other programming languages may use a *column-major* convention, in which a two-dimensional array is stored by columns rather than rows.

In Figure 12.1, the elements of A are shown in matrix form, and then as a linear sequence in the order in which they appear in memory. This suggests the interpretation of two-dimensional arrays as an array of one-dimensional arrays, each of which may be considered a row of a matrix. This, in fact, is the way two-dimensional arrays are usually viewed in C++. In particular, A[i] denotes the $(i + 1)st$ row of the array and may be used as a one-dimensional array, as will be done in the next section.

$$
\begin{array}{ccc}
\text{A}[0,0] & \text{A}[0,1] & \text{A}[0,2] \\
\text{A}[1,0] & \text{A}[1,1] & \text{A}[1,2] \\
\text{A}[2,0] & \text{A}[2,1] & \text{A}[2,2] \\
\\
\text{first row} & \text{second row} & \text{third row}
\end{array}
$$

| A[0, 0] A[0, 1] A[0, 2] | A[1, 0] A[1, 1] A[1, 2] | A[2, 0] A[2, 1] A[2, 2] |

Figure 12.1 Storage of Two-Dimensional Array

12.2 ARRAYS AND FUNCTIONS

One of the most basic operations in matrix algebra is multiplication of a vector by a matrix. If A is an $n \times n$ matrix and $\mathbf{x}$ is an n-long vector, the *matrix-vector product* $A\mathbf{x}$ is an n-long vector $\mathbf{b}$ whose components are defined by

$$
b_i = \sum_{j=1}^{n} a_{ij}x_j, \quad i = 1, \ldots, n. \tag{12.7}
$$

Inner Product

The *inner product* (or *dot product*) of two n-long vectors $\mathbf{u}$ and $\mathbf{v}$ is defined as

$$
(u, v) = \sum_{i=1}^{n} u_i v_i
$$

We may then interpret the matrix-vector multiplication (12.7) as

$$
b_i = (A_i, \mathbf{x}), \quad i = 1, \ldots, n \tag{12.8}
$$

where A_i is the ith row of A. The program in Example 12.1 carries out matrix-vector multiplication based on (12.8). This is illustrated for only 2×2 matrices, but the program is easily changed to be valid for other matrices also (Exercise 12.2).

In Example 12.1, a function `inprod()` is first defined to compute the inner product of two vectors. This function is then used in `main()` with the first array being `A[i]`; thus each row of the array `A` is accessed as a one-dimensional array. In Example 12.1 we are computing the matrix-vector product

$$
\begin{bmatrix} 3 & 2 \\ 3 & 4 \end{bmatrix} \begin{bmatrix} 4 \\ 6 \end{bmatrix} = \begin{bmatrix} 24 \\ 36 \end{bmatrix}
$$

and the components `b[0] = 24, b[1] = 36` will be printed by the `cout` statement.

In Example 12.2, a function for performing matrix-vector multiplication is given, again for the special case of 2×2 matrices. This function has a *nested* `for` loop, in which a `for` loop appears within another `for` loop. Such nesting of `for`

Nested `for` Loops

loops is common in matrix computations. The statement `b[i] = 0` ensures that this component of `b` is zero before beginning the following `for` loop; if the initial value of `b[i]` were not zero, the final result would be in error. If this `b[i] = 0` statement were not present, no braces would be required in this nested `for` loop.

EXAMPLE 12.1

Matrix-Vector Multiplication Based on Inner Products

```
#include <iostream.h>
float inprod(float u[ ], float v[ ], int n){
  float value = 0;              //This function
  for(int i = 0; i < n; i++) //computes
    value += u[i] * v[i];       //the inner product
  return value;
}
int main(){
  float A[2][2] = {{3., 2}, {3., 4}};
  float b[2], x[2] = {4., 6.};
  int n = 2;
  for(int i = 0; i < n; i++)   //This is the matrix-
    b[i] = inprod(A[i], x, n); //vector multiplication
  cout << "b[0] = " << b[0] << " b[1] = " << b[1] << endl;
  return 0;
}
```

Array Size Fixed

In the definition of the function `matvec()` of Example 12.2, as well as the function prototype, we have left the lengths of the arrays x and b blank, as well as the number of rows in A. However, it is necessary to give explicitly the number of columns in A; the reason for this will be discussed in the next subsection. In fact, the number of columns given in the function heading must be equal to the number of columns in the actual array parameter when the function is called. This is very restrictive, and in Chapters 15 and 16 we will see how pointers can be used to define a matrix-vector multiplication function in which the size of the arrays do not need to be fixed.

EXAMPLE 12.2

A Function for 2×2 Matrix-Vector Multiplication

```
#include <iostream.h>
void matvec(float [ ][2],float [ ],float [ ],int );
int main(){
  float A[2][2] = {{3.,2.},{3.,4.}};
  float b[2],x[2]={4.,6.};
  int n = 2;
  matvec(A,x,b,n);
  cout << "b[0] = " << b[0] << " b[1] = " << b[1];
  return 0;
}
void matvec(float A[ ][2],float x[ ],float b[ ],int n){
  for(int i = 0; i < n; i++){
    b[i] = 0;
    for(int j = 0; j < n; j++)
      b[i] = b[i] + A[i][j] * x[j];
  }
}
```

A Common Error

A possible pitfall in working with nested `for` loops is illustrated by the following statements:

```
for(int i = 0; i < n; i++){
   b[i] = 0;
   for(int j = 0; j < n; j++)
      b[i] = b[i] + A[i][j] * x[j];
}
for(int k = 0; k < n; k++)
   for(j = 0, j < n; j++) //j is undefined
      .
      .
      .
```

Here, j is defined as type `int` within the body of the `for` loop indexed by i. Hence, j is local to that `for` loop and must be redefined when it is used again.

The break *Statement*

Another caution with nested loops is the use of the `break` statement. If a `break` statement is executed in the inner loop of a nested loop, it will terminate only that inner loop. Thus, if your intent is to terminate both loops, you must have a `break` statement in the outer loop.

Inefficiencies in Arrays

Although two- and higher-dimensional arrays are very useful, there are some inefficiences associated with their use. Many matrices in applications have mostly zero elements. An important example is a *tridiagonal* matrix,

Tridiagonal Matrices

$$\begin{bmatrix} a_{11} & a_{12} & & & & \\ a_{21} & a_{22} & a_{23} & & & \\ & a_{32} & \ddots & & \ddots & \\ & & \ddots & & \ddots & a_{n-1,n} \\ & & & a_{n,n-1} & a_{nn} \end{bmatrix}$$

in which all elements are zero except those on the main diagonal and the two adjacent diagonals. If this matrix is stored in an $n \times n$ two-dimensional array, then n^2 storage locations will be reserved, even though only $3n - 2$ are required for the nonzero elements. Such matrices are usually stored in three one-dimensional arrays, one for each diagonal. In Part III, we will solve linear systems of equations with tridiagonal coefficient matrices stored in three one-dimensional arrays.

Another source of inefficiency is in accessing the elements of two- (and higher-) dimensional arrays. Suppose that we have the declaration

Accessing Array Elements

$$\text{float } A[50][100], \quad b[100]; \tag{12.9}$$

and the statement

$$y = A[i][j];$$

Since two-dimensional arrays are stored in memory one row after another, the compiler will set up code to compute the address of $A[i][j]$ as

$$BASE + (100 * i + j) * s \qquad (12.10)$$

Here BASE is the memory address of the first element of the array, s is the number of bytes in a floating point number, and 100 is the number of columns in the array, as given by (12.9). Thus (12.10) must be evaluated for the current values of i and j in order to give the location in memory of $A[i][j]$. However, for the statement

$$z = b[i];$$

where b is defined by (12.9), the corresponding calculation for the memory address is

$$BASE + i * s$$

which requires only an addition and a multiplication. Thus memory addresses of two-dimensional arrays are more expensive to compute. (However, if the address of $A[i][j]$ has been computed, the address of $A[i][j+1]$ may be obtained by adding s, and the address of $A[i+1][j]$ by adding s times the number of columns in the array. Many compilers will compute subsequent addresses in this more efficient way.) This discussion is not meant to imply that two-dimensional arrays should not be used, but only to indicate that their use may be more expensive than one-dimensional arrays.

The computation (12.10) to obtain the address of $A[i][j]$ illustrates why it is necessary to supply the number of columns of a two-dimensional array to a function. In the function matvec() of Example 12.2 the address of $A[i][j]$ must be computed within the function. The function call, matvec(A,b,x,n), supplies the beginning addresses of the arrays A, b, and x. However, for A the number of columns must also be supplied, as in done in the function heading.

12.3 ARRAYS OF STRINGS

In Chapter 11, we discussed how strings of characters could be stored in one-dimensional arrays. For many purposes, it is useful to also store *arrays of strings* and two-dimensional arrays are natural for this. The definition

```
char sentence[100][21]
```

defines a two-dimensional character array that we may interpret as 100 strings of 20 characters each plus the termination character. Thus we might use such an array to hold sentences containing up to 100 words, with each word a maximum of 20 characters.

Replacing a String   We now give an example of a typical task of text editors and word processors. We assume that we have a two-dimensional array s of 100 strings and we wish to replace each occurrence of *dog* by *cat*. The program to do this is given in Example 12.3. Recall from Chapter 11 that the cin operation stops reading a string when a blank is encountered. Thus, when the statement cin >> s[i] is executed, only nonblank characters (plus the termination character) are entered in s[i]. For each string s[i], the for j loop moves through the string comparing three successive characters to 'd', 'o', 'g'. If a match is found, these characters are replaced by 'c', 'a', and 't'.

The program in Example 12.3 has two serious deficiencies. First, the allocation of 20 characters to each word of the sentence can be very wasteful of storage since each word will usually contain far fewer than 20 characters. Second, replacing each character of the word one at a time is not very efficient or elegant. In Chapter 15 we will see a better way to handle problems of this kind by means of pointers.

EXAMPLE 12.3
Replacing a String

```
//This program reads up to 100 strings of up to 20 characters
//each and replaces every occurrence of dog by cat
#include <iostream.h>
int main(){
  char s[100][21];
  int n,i,j;
  cout << "how many words?\n";
  cin >> n;
  for(i = 0; i < n; i++){
    cout << "read a word\n";
    cin >> s[i];
    cout << s[i] << endl;
    for(j = 0; j <= 17; j++)
      if(s[i][j] == 'd' && s[i][j+1] == 'o' && s[i][j+2] =='g'){
        s[i][j] = 'c';
        s[i][j+1] = 'a';
        s[i][j+2] = 't';
      }
    cout << s[i] << endl;
  }
  return 0;
}
```

MAIN POINTS OF CHAPTER 12

- Two- and higher-dimensional arrays may be defined. Two-dimensional arrays are a natural way to store matrices, although the indices conflict with standard mathematical notation.

- Two-dimensional arrays are stored in row-major order, that is, one row after another.

- When a two-dimensional array is a parameter of a function, its second dimension must be given, and when the function is called, the argument must have the same second dimension.

- Two-dimensional arrays can be inefficient for storing matrices in which most of the matrix elements are zero. Also, access to elements of two-dimensional arrays is more costly than for one-dimensional arrays.

- Two-dimensional arrays of characters may be interpreted as arrays of character strings.

12.1 Let an array be declared by

```
float a[2][2] = {1.0, 2.0, 3.0, 4.0};
```

What is the value of y after the statements

```
float y = 0;
for(int i = 0; i < 2, i++)
    y = y + a[i][i];
```

12.2 Rewrite the programs of Examples 12.1 and 12.2 so that they can multiply 3-long vectors by 3×3 matrices. Test the programs for several matrices and vectors.

12.3 Write a function that will add two $m \times 4$ matrices A and B to produce a third matrix C as the sum. Write a function `main()` that will test your function. What changes to the function must you make so that it can add matrices of other sizes?

12.4 The Schur product C of two $m \times n$ matrices A and B is defined by

$$c_{ij} = a_{ij}b_{ij}, \quad i = 1, \ldots, m, \quad j = 1, \ldots, n.$$

Modify the function of Exercise 12.3 so that it computes the Schur product rather than the sum, and test it.

12.5 The product $C = AB$ of an $m \times n$ matrix A and an $n \times p$ matrix B is the $m \times p$ matrix C whose elements are

$$c_{ij} = \sum_{k=1}^{n} a_{ik}b_{kj}, \quad i = 1, \ldots, m, \quad j = 1, \ldots, p$$

Write a function that will compute the product of an $m \times 4$ and a 4×5 matrix and test your function. What changes to the function must you make to handle other sizes of matrices?

12.6 Rewrite the program of Example 12.3 so that it replaces any appearance of `help` by `save`. Can you modify it to replace any appearance of `help` by `saveme`?

12.7 Write a program that will read n test characters and put them in a one-dimensional character array. Then read in up to 100 words of maximum length

20 characters each and put them in a two-dimensional character array, one word per row. These words should be read from a file and the read should terminate on the end-of-file. Count the number of words that are read. Now declare a two-dimensional integer array whose i, j entry will be the number of occurrences of test letter j in word i. Output the test letters, the words and the counts of the test letters in the words.

13

LINEAR EQUATIONS

We now consider one of the most important problems in scientific and engineering computation: the solution of linear systems of equations. The main goals of this chapter are to present the Gaussian elimination method of solving linear systems, and to discuss its basic properties. We begin with an example of how a linear system arises.

13.1 LEAST-SQUARES APPROXIMATION

In many problems in science and engineering, we wish to approximate a set of data by a "best fit." As a very simple example, suppose that we have m measurements $w_1, \ldots, w_m$ of the weight of some object, perhaps obtained with m different scales. What shall we take as the "best" approximation, w, to the true weight, based on this data? The "principle of least-squares approximation" says we should minimize the sum of the squares of the deviations $w - w_i$; that is, we should minimize

$$g(w) = \sum_{i=1}^{m} (w - w_i)^2. \tag{13.1}$$

From calculus, we know that g takes on a relative minimum at a point w that satisfies $g'(w) = 0$ and $g''(w) \geq 0$. Since

$$g'(w) = 2 \sum_{i=1}^{m} (w - w_i), \quad g''(w) = 2m,$$

it follows that we can find the minimizing value by solving the equation

$$2 \sum_{i=1}^{m} (w - w_i) = 0.$$

This gives

$$w = \frac{1}{m} \sum_{i=1}^{m} w_i,$$

so that the best least-squares approximation is just the average of the measurements.

A Function of Three Variables

The same idea can be applied to obtaining functions that approximate given data. For example, suppose that we measure the temperature of some object as a function of time and we obtain temperatures $u_1, \ldots, u_m$ at the times $t_1, \ldots, t_m$. Suppose that we know (or assume) that the temperature should obey an equation of the form

$$u(t) = c_1 + c_2 t + c_3 \sin \pi t, \tag{13.2}$$

and we wish to estimate the unknown parameters $c_1, c_2,$ and c_3 from the measurements. The least-squares principle says that we then should minimize

$$g(c_1, c_2, c_3) = \sum_{i=1}^{m} (c_1 + c_2 t_i + c_3 \sin \pi t_i - u_i)^2 \tag{13.3}$$

as a function of $c_1, c_2,$ and c_3.

A Function of n Variables

This problem is a special case of the more general problem: Given measurements $u_1, \ldots, u_m$ at times $t_1, \ldots, t_m$, find a function of the form

$$u(t) = c_1 f_1(t) + c_2 f_2(t) + \cdots + c_n f_n(t) \tag{13.4}$$

that is the best least-squares approximation to the data. Here $f_1, \ldots, f_n$ are given functions; for example, for (13.2), $f_1(t) = 1$, $f_2(t) = t$, $f_3(t) = \sin \pi t$. Again, the parameters $c_1, \ldots, c_n$ in (13.4) are to be found by minimizing a sum of squares:

$$\sum_{i=1}^{m} (c_1 f_1(t_i) + c_2 f_2(t_i) + \cdots + c_n f_n(t_i) - u_i)^2. \tag{13.5}$$

The quantity to be minimized in (13.5) is a function of the n variables $c_1, \ldots, c_n$. Again, we can attack this problem by calculus, but now we must set partial derivatives to zero. We will not give the details of this development but just state the final result. In order to minimize (13.5) and obtain $c_1, \ldots, c_n$, we must solve the system of linear equations

$$\begin{bmatrix} a_{11} & a_{12} & \cdots & a_{1n} \\ a_{21} & & & \vdots \\ \vdots & & & \\ a_{n1} & \cdots & & a_{nn} \end{bmatrix} \begin{bmatrix} c_1 \\ c_2 \\ \vdots \\ c_n \end{bmatrix} = \begin{bmatrix} b_1 \\ b_2 \\ \vdots \\ b_n \end{bmatrix}, \tag{13.6}$$

where

$$a_{ij} = \sum_{k=1}^{m} f_i(t_k) f_j(t_k), \quad b_j = \sum_{k=1}^{m} f_j(t_k) u_k, \quad i, j = 1, \ldots, n. \tag{13.7}$$

Exercises 13.3 and 13.4 ask you to solve the system (13.6)/(13.7) in some particular cases, using the Gaussian elimination method to be discussed next.

13.2 GAUSSIAN ELIMINATION

In order to solve linear systems of equations, we will use the method of *Gaussian elimination*. We will first illustrate this method for the following system of 3 equations in 3 unknowns, in which the unknowns are now denoted by x_1, x_2, and x_3. The system

$$\begin{bmatrix} 4 & -9 & 2 \\ 2 & -4 & 4 \\ -1 & 2 & 2 \end{bmatrix} \begin{bmatrix} x_1 \\ x_2 \\ x_3 \end{bmatrix} = \begin{bmatrix} 2 \\ 3 \\ 1 \end{bmatrix} \tag{13.8}$$

may be written as

$$\begin{aligned} 4x_1 - 9x_2 + 2x_3 &= 2 \\ 2x_1 - 4x_2 + 4x_3 &= 3 \\ -x_1 + 2x_2 + 2x_3 &= 1 \end{aligned} \tag{13.9}$$

The first major step of Gaussian elimination is to eliminate the first variable, x_1, from the second and third equations. If we subtract 0.5 times the first equation from the second equation, and -0.25 times the first equation from the third equation, we obtain the equivalent system of equations:

$$\begin{aligned} 4x_1 - 9x_2 + 2x_3 &= 2 \\ 0.5x_2 + 3x_3 &= 2 \\ -0.25x_2 + 2.5x_3 &= 1.5 \end{aligned} \tag{13.10}$$

The second major step eliminates x_2 from the third equation. This is accomplished by subtracting -0.5 times the second equation of (13.10) from the third, leading to

The Triangular System

$$\begin{aligned} 4x_1 - 9x_2 + 2x_3 &= 2 \\ 0.5x_2 + 3x_3 &= 2 \\ 4x_3 &= 2.5 \end{aligned} \tag{13.11}$$

The second part of the algorithm consists of solving the *triangular* system of linear equations (13.11). This is easily accomplished by *back substitution*. The last equation of (13.11) is

Back Substitution

$$4x_3 = 2.5,$$

and, therefore, $x_3 = 2.5/4 = 0.625$. This value now is substituted into the second equation:

$$0.5x_2 + (3)(0.625) = 2;$$

hence, $x_2 = 0.25$. Substitution of x_2 and x_3 into the first equation yields

$$4x_1 - (9)(0.25) + (2)(0.625) = 2,$$

or $x_1 = 0.75$. To check this computed solution, we multiply

$$\begin{bmatrix} 4 & -9 & 2 \\ 2 & -4 & 4 \\ -1 & 2 & 2 \end{bmatrix} \begin{bmatrix} 0.75 \\ 0.25 \\ 0.625 \end{bmatrix}$$

which agrees with the right-hand side of (13.8).

Gaussian Elimination for n Equations

For a general $n \times n$ system, Gaussian elimination follows the same steps as for the 3×3 example. If the system is written in the form

$$
\begin{aligned}
a_{11}x_1 &+ \cdots + a_{1n}x_n &= b_1 \\
a_{21}x_1 &+ \cdots + a_{2n}x_n &= b_2 \\
&\vdots \\
a_{n1}x_1 &+ \cdots + a_{nn}x_n &= b_n
\end{aligned}
\tag{13.12}
$$

the first stage eliminates the coefficients of x_1 in the last $n-1$ equations by subtracting a_{21}/a_{11} times the first equation from the second equation, a_{31}/a_{11} times the first equation from the third equation, and so on. This gives the reduced system of equations

$$
\begin{aligned}
a_{11}x_1 + a_{12}x_2 &+ \cdots + a_{1n}x_n &= b_1 \\
a_{22}^{(1)}x_2 &+ \cdots + a_{2n}^{(1)}x_n &= b_2^{(1)} \\
&\vdots \\
a_{n2}^{(1)}x_2 &+ \cdots + a_{nn}^{(1)}x_n &= b_n^{(1)},
\end{aligned}
\tag{13.13}
$$

where

$$a_{ij}^{(1)} = a_{ij} - a_{1j}\frac{a_{i1}}{a_{11}}, \quad b_i^{(1)} = b_i - b_1\frac{a_{i1}}{a_{11}}, \quad i, j = 2, \ldots, n.$$

Precisely the same process is now applied to the last $n - 1$ equations of the system (13.13) to eliminate the coefficients of x_2 in the last $n - 2$ equations. We continue in this way until the entire system has been reduced to the *triangular system* of equations

Triangular System

$$
\begin{aligned}
a_{11}x_1 + a_{12}x_2 &+ \cdots + a_{1n}x_n &= b_1 \\
a_{22}^{(1)}x_2 &+ \cdots + a_{2n}^{(1)}x_n &= b_2^{(1)} \\
&\vdots \\
a_{nn}^{(n-1)}x_n &= b_n^{(n-1)}
\end{aligned}
\tag{13.14}
$$

where the superscripts indicate the number of times the elements have, in general, been changed. The solution of the triangular system (13.14) is now easily obtained by *back substitution*, in which the equations in (13.14) are solved in reverse order:

Back Substitution

$$x_n = \frac{b_n^{(n-1)}}{a_{nn}^{(n-1)}}$$

$$x_{n-1} = \frac{b_{n-1}^{(n-2)} - a_{n-1,n}^{(n-2)} x_n}{a_{n-1,n-1}^{(n-2)}} \qquad (13.15)$$

$$\vdots$$

$$x_1 = \frac{b_1 - a_{12}x_2 - \cdots - a_{1n}x_n}{a_{11}}$$

In the preceding, we have assumed that a_{11} and all the numbers $a_{ii}^{(i-1)}$, $i = 2, \ldots, n$, that are used as divisors are nonzero. We will return to this assumption shortly.

EXAMPLE 13.1
A Gaussian Elimination Function

```
//This function solves an n x n system of linear equations
//by Gaussian elimination.  The second dimension of the
//parameter A must be given and must match the array dimension
//of the actual parameter.
void gauss(float A[][4], float x[], float b[], int n){
  for(int k = 0; k < n; k++){
    float divisor = A[k][k];
    if(divisor == 0){ //check for zero divisor
      cout << "divisor is zero" << endl;
      return;
    }
    else //do triangular reduction
      for(int i = k + 1; i < n; i++){
        float multiplier = A[i][k]/divisor;
        for(int j = k + 1; j < n; j++)
          A[i][j] = A[i][j] -  multiplier * A[k][j];
        b[i] = b[i] - multiplier * b[k];
      }
  } //end of triangular reduction
  if(A[n-1][n-1] == 0){
    cout << "The last divisor is zero" << endl;
    return;
  } //do back substitution
  for(int i = n - 1; i >= 0; i--){
    x[i] = b[i];
    for(int j = i + 1; j < n; j++)
      x[i] = x[i] - A[i][j] * x[j];
    x[i] = x[i]/A[i][i];
  }
}
```

Pseudocode

The Gaussian elimination process is summarized in the pseudocode of Figure 13.1. This pseudocode is easily translated into a C++ function, as shown in Example

Figure 13.1 Pseudocode for Gaussian Elimination

Input: $n \times n$ matrix A, vector $\mathbf{b}$, n

Reduce system to triangular

For $k = 1, \ldots, n - 1$,

If $a_{kk} = 0$, error message and exit

For $i = k + 1, \ldots, n$

$$\ell_{ik} = \frac{a_{ik}}{a_{kk}}$$

For $j = k + 1, \ldots, n$

$$a_{ij} = a_{ij} - \ell_{ik} a_{kj}$$
$$b_i = b_i - \ell_{ik} b_k$$

Now solve the triangular system

If $a_{nn} = 0$, error message and exit

For $k = n, n - 1, \ldots, 1$

$$x_k = \frac{b_k - \sum_{j=k+1}^{n} a_{kj} x_j}{a_{kk}}$$

Output: Solution vector $\mathbf{x}$

C++ Function

13.1. The parameter A of the function gauss must have its second dimension given explicitly, as discussed in Chapter 12. This dimension must equal the corresponding dimension of the actual argument when the function is called. For example, in main() we might have the statements

```
int main(){
  const int n = 4;
  float A[n][n], b[n], x[n];
  //statements here to set values of A and b
  gauss(A, x, b, n);
    .
    .
    .
```

Here, the size of n, 4, matches the second dimension for A in the function of Example 13.1. But if the size of the system were different, say n = 6, this size would have to be used for the second dimension of A in Example 13.1, as well as in main(). In Chapter 16, we will see that by means of pointers we can change gauss() so that this restriction is removed.

Since the arrays are called by reference in Example 13.1, the original matrix A and vector b will be modified by the function gauss(). Thus, if you wish to retain the original copies of A and b, you must save them in other arrays before calling gauss(). Note that there are two returns in the function gauss, one if a divisor is zero and one if the (n, n) element of the triangular system is zero. In either case, after an error message is printed, control is returned to the calling program and the Gaussian elimination process is terminated.

Test Cases

Before using a program such as gauss(), it should be extensively tested; for this, one needs systems with known solutions. One easy way to obtain such systems is to choose a matrix A and a solution $\mathbf{x}$, and then form the corresponding $\mathbf{b}$. For example, if

$$A = \begin{bmatrix} 3 & 1 & 1 \\ 1 & 3 & 1 \\ 1 & 1 & 3 \end{bmatrix}, \quad \mathbf{x} = \begin{bmatrix} 1 \\ 2 \\ 3 \end{bmatrix}, \tag{13.16}$$

then

$$\mathbf{b} = A\mathbf{x} = \begin{bmatrix} 8 \\ 10 \\ 12 \end{bmatrix}. \tag{13.17}$$

Thus, if the vector $\mathbf{b}$ of (13.17) and the matrix A of (13.16) are given as inputs to the function gauss(), the solution vector $\mathbf{x}$ of (13.16) should be produced. Also, be sure you test for division by zero. A matrix that will cause division by zero for any vector $\mathbf{b}$ is

$$A = \begin{bmatrix} 1 & 1 & 1 \\ 1 & 1 & 1 \\ 1 & 1 & 1 \end{bmatrix}.$$

13.3 ERRORS

The Gaussian elimination program of Example 13.1 is suitable for least-squares problems because the matrices of such problems are very special. But for general linear systems of equations, it can give catastrophic errors. We now examine how this can happen.

Consider the system of linear equations

$$\begin{bmatrix} -10^{-5} & 1 \\ 2 & 1 \end{bmatrix} \begin{bmatrix} x_1 \\ x_2 \end{bmatrix} = \begin{bmatrix} 1 \\ 0 \end{bmatrix}, \tag{13.18}$$

whose exact solution is

$$x_1 = -0.4999975 \cdots, \qquad x_2 = 0.999995 \cdots.$$

Let us carry out Gaussian elimination on (13.18) using four-digit decimal arithmetic. The multiplier is

$$l_{21} = \frac{0.2 \times 10^1}{-0.1 \times 10^{-4}} = -0.2 \times 10^6,$$

which is exact, and the calculation for the new a_{22} is

Figure 13.2 Computed and Exact
Triangular Matrices

$$\begin{bmatrix} 10^{-5} & 1 \\ 0 & 0.2 \times 10^6 \end{bmatrix} \begin{bmatrix} 10^{-5} & 1 \\ 0 & 0.200001 \times 10^6 \end{bmatrix}$$

(a) Computed   (b) Exact

$$a_{22}^{(1)} = 0.1 \times 10^1 - (-0.2 \times 10^6)(0.1 \times 10^1)$$

$$= 0.1 \times 10^1 + 0.2 \times 10^6 \doteq 0.2 \times 10^6. \tag{13.19}$$

The exact sum in (13.19) is 0.200001×10^6, but since we are using only four digits, this must be represented as 0.2000×10^6; this is the first error in the calculation. Figure 13.2 shows the computed and exact triangular matrices.

The new b_2 is

$$b_2^{(1)} = -(-0.2 \times 10^6)(0.1 \times 10^1) = 0.2 \times 10^6. \tag{13.20}$$

No rounding errors occurred in this computation, nor do any occur in the back substitution:

$$x_2 = \frac{b_2^{(1)}}{a_{22}^{(1)}} = \frac{0.2 \times 10^6}{0.2 \times 10^6} = 0.1 \times 10^1,$$

$$x_1 = \frac{0.1 \times 10^1 - 0.1 \times 10^1}{-0.1 \times 10^{-4}} = 0.$$

Bad Error

The computed x_2 agrees excellently with the exact x_2, but the computed x_1 has no digits of accuracy. Note that the only error made in the calculation is in $a_{22}^{(1)}$, which has an error in the sixth decimal place. Every other operation was exact. How, then, can this one "small" error cause the computed x_1 to deviate so drastically from its exact value?

Note that the quantity 0.000001×10^6 that was dropped from the computed $a_{22}^{(1)}$ in (13.19) is the original element a_{22}. Since this is the only place that a_{22} enters the calculation, the computed solution would have been the same if a_{22} were zero. Therefore, the calculation using four digits has computed the exact solution of

*We Solved a Very
Different System*

the system

$$\begin{bmatrix} -10^{-5} & 1 \\ 2 & 0 \end{bmatrix} \begin{bmatrix} x_1 \\ x_2 \end{bmatrix} = \begin{bmatrix} 1 \\ 0 \end{bmatrix}. \tag{13.21}$$

Intuitively, we would expect the two systems (13.18) and (13.21) to have rather different solutions, and this is indeed the case. But why did this occur? The culprit is the large multiplier l_{21}, which made it impossible for a_{22} to be included in the sum in (13.19), using only four digits. This large multiplier was due to the smallness of a_{11} relative to a_{21}.

We can easily circumvent this problem by simply interchanging the order of the

Interchange Rows

equations:

$$\begin{bmatrix} 2 & 1 \\ -10^{-5} & 1 \end{bmatrix} \begin{bmatrix} x_1 \\ x_2 \end{bmatrix} = \begin{bmatrix} 0 \\ 1 \end{bmatrix}. \tag{13.22}$$

If we do Gaussian elimination on (13.22) using four digits, we obtain

$$l_{21} = \frac{-0.1 \times 10^{-4}}{0.2 \times 10^{1}} = -0.5 \times 10^{-5}$$

$$a_{22}^{(1)} = 0.1 \times 10^{1} - (-0.5 \times 10^{-5})(1) \doteq 0.1 \times 10^{1}$$

$$b_{2}^{(1)} = 0.1 \times 10^{1} - (-0.5 \times 10^{-5})(0) = 0.1 \times 10^{1}$$

$$x_{2} = \frac{0.1 \times 10^{1}}{0.1 \times 10^{1}} = 1.0$$

$$x_{1} = \frac{-(0.1 \times 10^{1})(1)}{0.2 \times 10^{1}} = -0.5.$$

The computed solution now agrees excellently with the exact solution.

Partial Pivoting

This example illustrates that the Gaussian elimination algorithm may give inaccurate results if the multipliers are large. The example also indicates that this difficulty might be remedied by interchanging equations; this is in fact the case. By a relatively simple strategy we can always arrange to keep the multipliers in the elimination process less than or equal to 1 in absolute value. This is known as *partial pivoting*: At the kth stage of the elimination process an interchange of rows is made, if necessary, to place in the main diagonal position the element of largest absolute value from the kth column on or below the main diagonal. Figure 13.3 gives this modification to the part of the pseudocode of Figure 13.1 that reduces the original system to triangular form. The back substitution remains the same.

Gaussian elimination with partial pivoting has proved to be an extremely reliable algorithm in practice. However, the matrix must be properly scaled before the algorithm is used. To illustrate this, consider the system

$$\begin{bmatrix} 10 & -10^{6} \\ 2 & 1 \end{bmatrix} \begin{bmatrix} x_{1} \\ x_{2} \end{bmatrix} = \begin{bmatrix} -10^{6} \\ 0 \end{bmatrix}, \tag{13.23}$$

which is just the original system (13.18) with the first equation multiplied by -10^{6}. (Multiplication of an equation by a constant does not change the solution.) No

Figure 13.3 Reduction to Triangular System with Partial Pivoting

For $k = 1, \ldots, n - 1$

    Find $m \geq k$ such that $|a_{mk}| = \max\{|a_{ik}| : i \geq k\}$.

    If $a_{mk} = 0$, then error message and exit.

    Else interchange a_{kj} and a_{mj}, $j = k, k + 1, \ldots, n$.

        interchange b_{k} and b_{m}.

    For $i = k + 1, k + 2, \ldots, n$

    $l_{ik} = a_{ik}/a_{kk}$

    For $j = k + 1, k + 2, \ldots, n$

        $a_{ij} = a_{ij} - l_{ik}a_{kj}$

        $b_{i} = b_{i} - l_{ik}b_{k}.$

Scaling

interchange is called for by the partial pivoting strategy since the $(1, 1)$ element is already the largest in the first column. However, if we carry out the elimination using four digits (see Exercise 13.6), we will encounter exactly the same problem that we did with the system (13.18). To avoid this problem, rows of the matrix should be scaled so that the maximum absolute value of the elements in each row is approximately 1. If we do this with (13.23), the $(1, 1)$ element will then become small, and the partial pivoting strategy will cause an interchange of the equations.

Ill-Conditioned Linear Systems

We next give another example that at first glance looks like simple rounding error, but there is a deeper root cause. Consider the system

$$
\begin{aligned}
0.832x_1 + 0.448x_2 &= 1.00 \\
0.784x_1 + 0.421x_2 &= 0,
\end{aligned}
\tag{13.24}
$$

and assume that we now use three-digit decimal arithmetic to carry out Gaussian elimination. Since a_{11} is the largest element of the matrix no interchange is required, and the computation of the new elements $a_{22}^{(1)}$ and $b_1^{(1)}$ is

$$
l_{21} = \frac{0.784}{0.832} = 0.942\underline{308} \cdots \doteq 0.942
$$

$$
a_{22}^{(1)} = 0.421 - 0.942 \times 0.448 = 0.421 - 0.422\underline{016} \doteq -0.001 \tag{13.25}
$$

$$
b_2^{(1)} = 0 - 1.00 \times 0.942 = -0.942,
$$

where the underscored digits are lost in the three-digit computation. Hence, the computed triangular system is

$$
\begin{aligned}
0.832x_1 + 0.448x_2 &= 1.00 \\
-0.001x_2 &= -0.942,
\end{aligned}
\tag{13.26}
$$

and the back substitution produces the approximate solution

$$
x_1 = -506, \qquad x_2 = 942. \tag{13.27}
$$

More Errors

But the exact solution of (13.24), correct to three figures, is

$$
x_1 = -439, \qquad x_2 = 817, \tag{13.28}
$$

so the computed solution is incorrect by about 15%. Why has this occurred?

The first easy answer is that we have lost significance in the calculation of $a_{22}^{(1)}$. Indeed, it is clear that the computed value of $a_{22}^{(1)}$ has only one significant figure; so our final computed solution will have no more than one significant figure. But this is only the manifestation of the real problem. By carrying out a more detailed computation, we can show that the computed solution (13.27) is the exact solution of the system

What We Really Solved

$$0.832x_1 + 0.447974\cdots x_2 = 1.00$$
$$0.783744\cdots x_1 + 0.420992\cdots x_2 = 0,$$

(13.29)

where $\cdots$ in the coefficients indicates that only 6 figures are shown. The maximum percentage change between the elements of this system and the original system (13.24) is only 0.03%; therefore, errors in the data are magnified by a factor of about 500.

Ill-Conditioning

Whenever a problem is such that small changes in the data of the problem can cause large changes in the solution, the problem is called *ill-conditioned*. In the case of (13.24) small changes (0.03%) in the coefficients of the matrix cause large changes (15%) in the solution. Thus, just because the coefficients in the matrix are accurate to three decimal places does not mean one can expect a solution with similar accuracy. The fault is *not* that of the Gaussian elimination algorithm but of the problem itself.

Geometry of Ill-Conditioning

We can understand geometrically the cause of this ill-conditioning in (13.24). Figure 13.4 shows a plot of the two equations of (13.24). The intersection (not shown) of these two lines is the solution of the system. As shown in Figure 13.4 the lines are almost parallel, and it is this that causes the ill-conditioning since very small changes in the slopes of the lines can cause very large changes in their intersection point. In fact, consider the system of equations

$$0.832x_1 + 0.448x_2 = 1.00$$
$$0.784x_1 + (0.421 + \varepsilon)x_2 = 0.$$

(13.30)

The second equation defines a family of lines depending on the parameter ε. As ε increases from zero to approximately 0.0012, the line rotates counterclockwise, and its intersection with the line defined by the first equation recedes to infinity until the two lines become exactly parallel and no solution of the linear system exists.

Ill-conditioned problems pervade all areas of scientific and engineering compu-

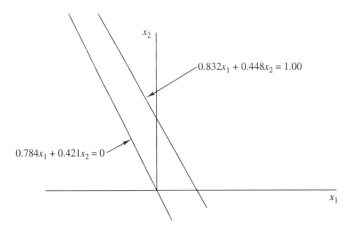

Figure 13.4 Almost-Parallel Lines Defined by (13.24).

tations, not just linear equations, and it is necessary to be alert to such possibilities when it appears that an algorithm might be giving errors.

13.4 EFFICIENCY

We have previously discussed some inefficiencies in using certain C++ constructs such as two-dimensional arrays. But these inefficiencies are minor compared to inefficiencies that can result from the use of poor algorithms. We next give an example of how a very inefficient method could arise in the context of linear equations.

Computing Determinants

Many elementary textbooks on linear algebra present *Cramer's rule* as a method for solving linear systems of equations. This method involves computing quotients of certain determinants of matrices. For 2×2 and 3×3 matrices A, the determinant is given by the formulas

$$\det A = a_{11}a_{22} - a_{12}a_{21}, \quad n = 2$$

$$\det A = a_{11}a_{22}a_{33} + a_{12}a_{23}a_{31} + a_{13}a_{21}a_{32}$$
$$- a_{11}a_{23}a_{32} - a_{13}a_{22}a_{31} - a_{12}a_{21}a_{33}, \quad n = 3$$

A Very Inefficient Algorithm

For general n, these simple formulas generalize to the sum of all possible products (half with minus signs) of elements of the matrix, one element from each row and each column. There are $n!$ such products for an $n \times n$ matrix. (In particular, 2 for $n = 2$ and 6 for $n = 3$, as shown above.) Now, if we proceed to carry out the computation of a determinant based on a straightforward implementation of this formula, it would require about $n!$ multiplications and additions. For n very small, say $n = 2$ or 3, this is a small amount of work. Suppose, however, that we have a 20×20 matrix, a very small size in current scientific computing. If we assume that each arithmetic operation requires 1 microsecond (10^{-6} second), about the speed of a slow PC, then the time required for this calculation—even ignoring all overhead operations in the computer program—will exceed one million years! On the other hand, the Gaussian elimination method will do the arithmetic operations for the solution of a 20×20 linear system in less than 0.005 second, again assuming 1 microsecond per operation. Although this is an extreme example, it does illustrate the difficulties that can occur by naively following a mathematical prescription in order to solve a problem on a computer.

Faster Than Gaussian Elimination?

In general, we would like to choose a method that minimizes the computing time yet retains suitable accuracy. For some relatively simple problems, an estimate of the computing time may be based on counting the number of arithmetic operations required. For example, for Gaussian elimination, approximately $\frac{1}{3}n^3$ multiplications and $\frac{1}{3}n^3$ additions are required for large n. (Note the tremendous difference in how rapidly $n!$ grows as compared to n^3: for $n = 20$, $n^3 \doteq 10^4$ but $n! \doteq 10^{18}$.) Is this the best that can be done? For many years it was thought that Gaussian elimination was probably the optimal algorithm, but in 1969 a method was developed in which the number of multiplications was proportional to $n^{2.81}$, which is considerably smaller than n^3 for large n. However, this method is much more complicated than Gaussian elimination; only for very large n (for example, $n = 1000$) is it actually faster.

The study of the efficiency of algorithms has led to the subarea of computer science called *computational complexity*. However, only for relatively simple problems is it possible to ascertain very precisely how fast a given algorithm will be. Usually, timing comparisons of competitive algorithms must be made in order to determine the most efficient one.

MAIN POINTS OF CHAPTER 13

- Least-squares approximation leads to a linear system of equations that can be solved by Gaussian elimination without row interchanges.
- Gaussian elimination may be incorporated into a function and used as part of a larger program.
- In general, Gaussian elimination can be a very inaccurate algorithm unless partial pivoting and scaling are incorporated.
- Even with a good algorithm, solution of linear systems may be inaccurate if the system is ill-conditioned.

EXERCISES

13.1 Solve the linear systems

$$\begin{bmatrix} 1 & 2 \\ 3 & 4 \end{bmatrix} \begin{bmatrix} x_1 \\ x_2 \end{bmatrix} = \begin{bmatrix} 1 \\ 2 \end{bmatrix} \qquad \begin{bmatrix} 1 & 2 & 3 \\ 4 & 5 & 6 \\ 7 & 8 & 7 \end{bmatrix} \begin{bmatrix} x_1 \\ x_2 \\ x_3 \end{bmatrix} = \begin{bmatrix} 1 \\ 2 \\ 3 \end{bmatrix}$$

by Gaussian elimination using a hand calculator. Delineate clearly the two major steps: reduction to a triangular system and back substitution to solve this triangular system.

13.2 Write a function `main()` that reads from the keyboard a matrix A and vector b and then calls the function `gauss()` of Example 13.1 (after changing the column dimension in the parameter A) to solve the system `Ax = b`. Test your program on the systems of Exercise 13.1 and also on the systems

$$\begin{bmatrix} 0 & 1 \\ 1 & 2 \end{bmatrix} \begin{bmatrix} x_1 \\ x_2 \end{bmatrix} = \begin{bmatrix} 1 \\ 2 \end{bmatrix} \qquad \begin{bmatrix} 1 & 1 \\ 1 & 1 \end{bmatrix} \begin{bmatrix} x_1 \\ x_2 \end{bmatrix} = \begin{bmatrix} 1 \\ 2 \end{bmatrix}$$

Discuss the behavior of `gauss()` on these last two systems.

13.3 Write a function `main()` to carry out the least squares algorithm summarized by (13.6) and (13.7). Restrict your attention to `n = 3`, and assume that the functions f_1, f_2, and f_3 will be given by function programs. Use the function `gauss()` of Example 13.1 to solve the linear system. Test your program on the problem $n = 2$, $f_1(t) = 1$, $f_2(t) = t$ [i.e., the function of (13.2) with $c_3 = 0$] and the data $t_1 = 0$, $t_2 = \frac{1}{2}$, $t_3 = 1$, $u_1 = 0$, $u_2 = 1$, $u_3 = 3$.

13.4 Apply your program of Exercise 13.3 to obtain the least-squares function (13.2) for the data

$$t_1 = 0, t_2 = \frac{1}{4}, t_3 = \frac{1}{2}, t_4 = \frac{3}{4}; \qquad u_1 = 1.5, u_2 = 4.0, u_3 = 5.5, u_4 = 5.0.$$

If you have access to a graphics system, plot the input data and your approximating least-squares function.

13.5 Modify the Gaussian elimination function of Example 13.1 so as to incorporate partial pivoting as given in the pseudocode of Figure 13.3. Test this new function on the systems of Exercise 13.2.

13.6 Using four-digit decimal arithmetic, as done with the example (13.18), show that Gaussian elimination fails for the system (13.23) in the same way it does for (13.18).

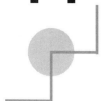

14

MORE ON FUNCTIONS

In Chapter 5 we introduced functions, and we have since defined and used several of them. In this chapter, we collect a number of other facts about functions, and give several additional examples.

14.1 REFERENCE VERSUS VALUE

Recall the function of Example 9.6 to obtain the deriative of a polynomial and reproduced as Example 14.1.

EXAMPLE 14.1
A Function to Obtain a Derivative

```
void p_prime(float a[ ], float der[ ], int n){
  for(int i = 1; i <= n; i++)
    der[i - 1] = i * a[i];
}
```

Suppose this function is called in main() by the statement

$$p\_prime(poly, der\_poly, m); \qquad (14.1)$$

where poly and der_poly have been declared as arrays. As discussed in Chapter 9, the current value of m is passed to the function, and a temporary memory location is created by the function to hold this value. This is *call by value*. The arrays poly and der_poly, however, are *called by reference*: The beginning address of each array is passed to the function. The address of poly replaces a in the function so that elements of the array poly are obtained directly from their storage positions as defined in main(). Similarly, the address of der_poly replaces der in the function, and the results of the computation are stored directly in the storage locations of der_poly as defined in main().

Side Effects

A *side effect* (so-called for historical reasons) occurs when a value in a calling program is changed by a function. In most cases, a side effect is part of the design of the function and is intended to occur, as with the array der in the previous example. However, as a result of an error in the function, unwanted side effects can also occur and damage data in the calling function. Suppose, for example, we made a mistake in the program of Example 14.1 and wrote the statement in the for loop as

$$a[i] = i * der[i - 1]; \qquad (14.2)$$

The effect of the for loop now would be to replace the elements of a. This kind of unwanted side effect is a danger with call by reference, but it can be prevented by requiring that the input data be const. Thus, if we change the function heading in Example 14.1 to

const *Parameters*

```
float p_prime(const float a[ ], float der[ ], int n)
```

then a is again called by reference, but no changes in this array are permitted in the function. A mistake such as (14.2) would now be caught during compilation.

If the argument of a function is const and within the function another function is called with the same argument, then this argument must also be const in the second function. For example, if the function called another function with the heading

```
float swap(int m, const float a[ ])
```

it would be an error to omit the const in the definition of this function since a must not be changed. If this function changed the array a, then it would also be changed in the calling function, which is not allowed.

A side effect cannot happen with call by value. Thus, if we had the following statement in the function of Example 14.1:

```
n = n + 2;
```

and the function was called by the statement (14.1), the value of m in the calling program would not be changed: n is a local variable in the function. Thus the use of const int n in the function definition is redundant as far as preventing unwanted side effects, but it does prevent any change of n in the function.

Reference Declarator

It is also possible to call single variables by reference. But to do this we need the mechanism illustrated in Example 14.2. The key thing in Example 14.2 is the use of the *reference declarator* & in the definition of the function example(). This causes the arguments of the function to be the address of an int variable j and the address of a float variable y. The function prototype is consistent with the definition, indicating that the arguments are to be addresses. Then, when the function is invoked in main(), the addresses of i and x, not their current values, are passed to the function. Thus the statement y = y * y; in the function replaces x in main() by 2.0 * 2.0; likewise i in main() is replaced by 3 + 3. The cout statement then displays the values

$$i = 6, x = 4$$

EXAMPLE 14.2
A Function with Call by Reference

```
#include <iostream.h>
void example(int&, float&); //prototype
int main(){
  int i = 3;
  float x = 2;
  example(i, x);
  cout << " i = " << i << ", x = " << x << endl;
  return 0;
}
void example(int &j, float &y){
  y = y * y;
  j = j + j;
}
```

The variables j and y in the function example() are called *reference parameters* since they are called by reference. The function heading in Example 14.2 may also be written as

```
void example(int& i, float& y){
```

and this placement of & is preferred by some authors.

The function example() of Example 14.2 is designed so that the values of i and x in main() are changed when the function is called. In other cases, a variable may be called by reference but with no intent to change its value in main(). To prevent any unwanted side effects with such variables, we could use, as with arrays, a function heading such as

const *Reference Parameters*

```
void func(int &k, const int &m);
```

Here, const prevents the variable m from being changed in the function. For example, a statement such as m = m + 1 within the function would cause a syntax error. In general, whenever we wish to ensure that variables called by reference cannot be changed, we should use const. The absence of const is an indicator that a variable *is* to be changed.

Functions May Return References

It is also possible for a function to return a reference rather than a value. This is accomplished by using & before the function name in the definition, as illustrated below:

```
float& maximum(float a[ ], int n){
  // statements here
  return a[m];
}
```

For example, this function might compute the maximum element, a[m], of the first n elements of the array a. Since & appears before the function name, the return value will be a reference, in this case the address of a[m].

When a function returns a reference, it is possible to use the function on the left side of an assignment statement, as in maximum(a, 5) = 0;. Here the function call produces the address of the maximum element in the first five elements of the array a, and the assignment statement then replaces this maximum element by 0. This assignment statement would not be possible unless the function returned a reference. (See Exercise 14.11.)

*Shifting Array
Indices*

As another example of returning a reference, consider the function

```
float& index(float a[ ], int i){
  return a[i - 1];
}
```

Given the array a and index i as parameters, this function will return the address of a[i - 1]. Thus the effect is to shift the array index, as in:

```
float a[3], x[3] = {1, 2, 3};
for(int i = 1; i <= 3; i++)
  index(a,i) = index(x, i);
```

Note that i runs from 1 to 3 in the for loop, which assigns the elements of the array x to the array a. [Here, the address given by index(x, i) is "dereferenced" since it appears after the = sign, and the value of element i - 1 of x is assigned, not the address of this element. Dereferencing will be discussed in detail in Section 15.1.] The notation index(x, i) is rather clumsy, and we will see in Part III a way to shift array indices but still keep the usual array notation x[i].

*Do Not Return
Local Variables by
Reference*

Warning: *Do not return local variables by reference.* For example, if we modified the previous function max to

```
float& max(float a[ ], int n){
  float value;
    //statements
  return value;
}
```

this would return a reference to a local variable, value, that doesn't exist outside the function.

14.2 RECURSIVE AND INLINE FUNCTIONS

It is possible for a function to call itself, in which case the function is called *recursive*. In Example 14.3 we give the standard example of a recursive function: the factorial function n != n * (n - 1) * (n - 2) ... * 2 * 1.

EXAMPLE 14.3

A Recursive Factorial Function

```
int fac(int n){
  int value;
  if(n == 1)
    value = 1;
  else
    value = n * fac(n - 1); //This is the recursive call
  return value;
}
```

Suppose that n = 3 when the function fac() is first called. Then the else part of the if statement is activated and fac() is called again to compute fac(2). This in turn calls fac() once more to compute fac(1), which returns the value 1. The call of fac(2) is now complete and returns the value 2, which is multiplied by 3 to give the final value of 6 for fac(3).

EXAMPLE 14.4

A Nonrecursive Factorial Function

```
int factorial(int n){
  int value = 1;
  for(int i = 2; i <= n; i++)
    value = value * i;
  return value;
}
```

Recursive functions may be elegant, but they can also be inefficient since they require keeping all data used by the function while the recursion proceeds. But good compilers can help to minimize this overhead. On the other hand, it is always possible to write a recursive function in a nonrecursive way. For example, Example 14.4 gives a nonrecursive factorial function. Other recursive functions may be much more difficult to convert to nonrecursive form, however.

Poor Use of Functions

As discussed in Chapter 5, when a function is called, a transfer is made from the calling program to the function, and values of actual parameters and reference addresses must be passed to the function. Other overhead is also incurred. Thus there is a cost involved in invoking a function, and if the function does not do substantial work, this cost can be relatively significant. However, it is rarely necessary to worry about the loss of efficiency from using functions except in critical parts of the program, the so-called "inner loops." (A general rule of thumb that applies to many large programs is that over 90% of the time is consumed by fewer than 10% of the statements.)

We illustrate this by the code for matrix-vector multiplication given in Example 12.2:

```
for(int i = 0; i < n; i++){
  b[i] = 0;
  for(int j = 0; j < n; j++)
    b[i] = b[i] + A[i][j] * x[j];
}
```

As an extreme example of the poor use of a function, suppose we defined a function by

```
float axpy(float x, float y, float z){
  return x + y * z;
}
```

We could then write the last statement of the matrix-vector multiplication code as

```
b[i] = axpy(b[i], A[i][j], x[j]);
```

Since it is this statement that is repeated over and over (n^2 times), we would incur a severe efficiency penalty by using a function at this point (unless it were inline, see below). Moreover, it would be rather absurd to use a function for this simple calculation, since this would tend to obscure the program.

Inline Functions

It is possible to avoid the overhead cost of calling a function by putting the function *inline*. This means that the code for the function, rather than a call to the function, is placed in the calling program where the function is needed. Then there is no need to pass parameters to a function or transfer to another section of code. If, for some reason, someone really wanted to write the above matrix-vector multiplication code in terms of the function axpy, a request could be made to the C++ compiler to put the function inline by writing inline before the function definition:

```
inline float axpy(float x, float y, float z){
  return x + y * z;
}
```

Such an inline function *must* be defined before the program that will use it. Thus, if the function were to be used in main(), the above definition must appear before main(). No prototypes are used with inline functions, since their complete definition must be given. (Inline functions in classes, to be discussed in Part III, are an exception.)

The keyword inline in the function definition is a *request* to the compiler to put the function inline, and there is no guarantee that the compiler will do this. It is always possible for the programmer to write out the function inline herself. But if a function is used in many different places in a program, this could become very tedious. Moreover, writing out the function code explicitly everywhere it is used is equivalent to not having a function at all, thus eliminating the benefits of modularity and reuse that a function provides. Generally, the compiler will honor the request to put a function inline, unless the function contains loops or is too large or complicated.

14.3 FUNCTION MISCELLANEA

In this section we collect a number of additional properties of functions.

Function Overloading

It is possible to have different functions with the same name. Suppose we have a program that works with vectors with both 2 and 3 components, and needs to compute the lengths of such vectors by the formulas:

$$\text{length}(x, y) = (x^2 + y^2)^{\frac{1}{2}} \tag{14.3a}$$

$$\text{length}(x, y, z) = (x^2 + y^2 + z^2)^{\frac{1}{2}} \tag{14.3b}$$

Two Functions with Same Name
We can define the functions to do these computations by:

```
float length(float x, float y){
  return sqrt(x * x + y * y);
}
float length(float x, float y, float z){
  return sqrt(x * x + y * y + z * z);
}
```

(These simple functions are good candidates to be declared inline.) Note that both functions have the same name, `length`, but we can define and use both functions in the same program. We say that we have *overloaded* the function name so that it has different meanings depending on the context of usage.

In the calling program, we might have statements such as

```
u = length(a,b);
v = length(r, s, t);
```

The compiler will know which function `length` to call by the number of arguments. Note that the function with return type `double`:

```
double length(float x, float y){
  return sqrt(x * x + y * y);
}
```

is *not* a correct overloading of the original `length` function since an overloading is determined by the number and types of the arguments and not the return type.

Poor Overloading
It is also possible to overload functions in unsatisfactory ways. For example,

```
float length(float x, float y, float z){
  return x * y * z;
}
```

would be a legal overloading of the first `length` function but would be very poor practice since this function has nothing to do with length.

Default Parameters

Many functions in scientific computing include arguments that are rarely used, and it would be nice to not have to specify these arguments every time the function is called. For example, suppose that

```
float integrate(float a, float b, float epsilon);
```

is the prototype for a numerical integration function over the interval (a,b). The parameter epsilon requests a certain accuracy, and a naive user of this function may not know how to choose it. It would be better to have the function use a default value of epsilon unless the user wishes to override that default. However, if we try to call the function by

$$y = integrate(0,1); \qquad (14.4)$$

the compiler will issue a syntax error since the call of the function has one fewer argument than the function definition.

A Default
Parameter

We can circumvent this problem by the use of *default* parameters. In this case, arguments that may be omitted in the function call are given default values in the function definition. For example, the first line of the definition of the function integrate might be

```
float integrate(float a, float b, float epsilon = .000001)
```

Now, a call of the form (14.4) is legal, and the default value of .000001 is used for epsilon in the function. Also, a call of the form

```
y = integrate(0, 1, .0001);
```

is legal, and the function now uses .0001 for epsilon in place of the default value.

Default Parameters
Must Go Last

Any number of default arguments may be used, but they must always appear after the nondefault arguments. For example, the following specification is not correct:

```
float integrate(float epsilon = .000001, float a, float b)//error
```

If two or more default arguments are used, arguments in the calling statement may be dropped only right to left. For example, if the function specification is

```
float f(int n, float a, float b = 1.0, float c = 2.0)
```

a call of the form

```
f(10, 1.0)
```

would use the default values of b and c. A call of the form

```
f(10, 1.0, 2.0)
```

would set b = 2.0 and use the default value of c. But we are not able to use the default value of b and set a new value of c unless we explicitly write in the default value of b:

```
f(0, 1.0, 1.0, 1.0)
```

Thus default arguments should be listed in the function heading so that the ones most likely to be changed come first.

Don't Give Default Parameters Twice

Default parameters may be given either in the function prototype or the function definition, but not both, even if the parameter values are the same. Thus, if the prototype is

```
float f(int n, float a, float b = 1.0); //prototype
```

the function definition should begin with

```
float f(int n, float a, float b) { //ok
```

and not

```
float f(int n, float a, float b = 1.0) { //error
```

It is best to give default parameters in the prototype so that the compiler knows about these even if it hasn't seen the function definition.

Static Variables

Suppose that the beginning lines of a function definition are

```
int func(int a) {
    int n;
```

Auto Variables

Then storage for a and n is automatically allocated each time the function is called. Local variables and parameters called by value are called *automatic* or *auto* variables, since they are automatically created when the function is called and automatically destroyed when the function is terminated. To make this explicit, we could write the declaration of n in the form:

```
auto int n; //rarely seen
```

Here, auto indicates that storage is to be automatically defined for n. But this is redundant, and the key word auto is rarely used.

Each time the function is called, new storage is defined for auto variables, and these storage locations may be different than on previous calls of the function. In particular, local variables do not usually retain their values from the previous time the function was called, and they must be assumed to have unknown values until set by the current function call. Sometimes, however, it is desirable to be able to save values of local variables between function calls; this may be achieved by declaring a variable to be *static*.

For example, suppose the beginning statements of a function definition are

```
int func(int a) {
    static int n = 1;
    n++;
```

Static Variables Are Unchanged

The definition of n as a static variable does two things: First, the value of n is initialized to 1 the first time the function is called, but only the first time. (If no explicit initialization is given, a static variable is initialized to zero.) Second, the storage location for n is defined only the first time the function is called, and any subsequent calls of the function use the same storage location for n. Thus static variables are like global variables: Their storage location is set once and for all. The result of this is that when the function is called a second time, the value of n is exactly what it was at the end of the previous call of the function. As a possible use of this capability, the variable n might count the number of times the function has been called on this run of the overall program.

Templates

Templates are a way to generate automatically different versions of functions that differ in the types of their arguments and return values. Example 14.5 shows a template for a function that finds the maximum of two parameters.

EXAMPLE 14.5
A Template for Maximization Functions

```
template <class T> T max(const T &a, const T &b){
  if(a < b)
    return b;
  else
    return a;
}
```

In Example 14.5, the *template argument list* is in brackets following the key word `template`. `T` is called a *metaparameter* that appears in the argument list as well as before the function name and before the function parameters. Then the function definition follows. Once the template is given, it can be used in the following ways:

Automatic Function Generation

```
int i, j
cin >> i >> j;
cout << max(i,j);
```

or

```
float x = 10.0, y = 2.0;
cout << max(x, y);
```

In the first case, a function `max` is generated that accepts integer arguments and returns an integer result. Here, the compiler recognizes that the arguments of `max` are of type `int` so the metaparameter `T` in Example 14.5 is set to `int` and the function `max` generated. In the second case the compiler recognizes that the arguments of `max` are of type `float` and generates a function `max` that accepts `float` arguments as input and returns a `float` result. However, the following is incorrect:

```
float x = 10.0;
int i = 20;
cout << max(i, x); //illegal
```

since the arguments of the function must be the same type. In using templates, there are no automatic type conversions.

The template definition must be placed before `main()`, or at least before a function that requires it is used. We will see in Part III that we may also use templates to generate classes in a similar way.

14.4 LIBRARIES OF FUNCTIONS

Header Files

We have used several libraries of functions: `math.h`, `iostream.h`, `stdio.h`, `iomanip.h`, and so on. (There are still other standard libraries, but they are not particularly useful for scientific computing.) The `.h` suffix indicates that this is a *header file* (also called an *interface* file). This file does not contain the actual programs for the functions but only the prototypes. The prototype, however, contains all the information that the compiler needs to set up a call to the function. Thus, if a program has the statements

```
float f(float, float); //prototype
int main(){
  //other statements here
  y = f(10, x);
```

the compiler will know that `f` is a function of type `float` with two arguments of type `float` and will be able to compile the statement `y = f(10,x);`. Without the prototype before `main()`, `f` would be undefined, and the compiler will issue an error message. The same is true when you use a standard mathematical function such as `sqrt()`. The directive

```
#include <math.h>
```

before `main()` furnishes the compiler with the prototype for `sqrt()`; without this prototype, `sqrt()` would be undefined, and, again, you would receive an error message.

Linking

As mentioned, header files such as `math.h` do not contain the actual programs for the functions. These programs are contained in another file and are attached to your program by a process called *linking*. The idea is illustrated in Figure 14.1.

Figure 14.1 A Schematic of Compiling and Linking

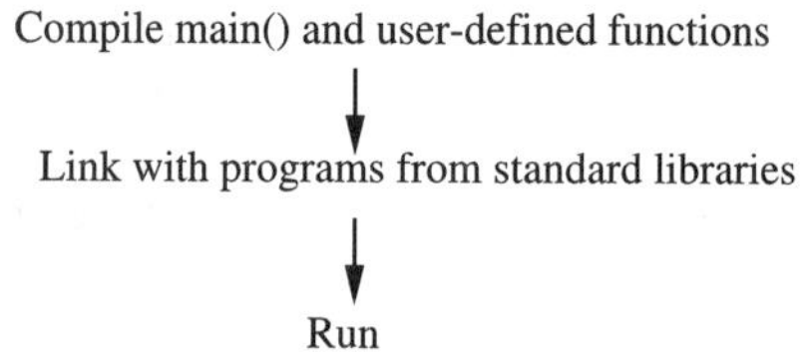

Compile main() and user-defined functions

↓

Link with programs from standard libraries

↓

Run

Constructing a Header File

In all the programs discussed so far that had user-defined functions, these functions were either defined before `main()`, or a prototype was put before `main()` and the functions defined later. In either case, the functions were part of the program file that is compiled, as indicated in Figure 14.1. Alternatively, we can construct our own libraries of functions (and other entities such as classes, to be discussed in Part III). First, a header file would be constructed that would contain the prototypes of all the functions in our own library. It could also contain other things such as global constants. Such a header file is illustrated in Example 14.6.

EXAMPLE 14.6
A Header File

```
//Header File
//Author: J. Student
//Date: 7/19/97
//Purpose: Define pi as global constant and
//give function prototypes
const float PI = 3.14 //Global Constant
float f(float, float); //Function prototype
int min(int, int); //Function prototype
```

Suppose that the name of the header file is `MyFunc.h`. Then before `main()` we would put a preprocessor directive of the form:

$$\text{\#include "A:MyDir\textbackslash MyFunc.h"} \tag{14.5}$$

Here, between the quotes we give the path of the file: the file `MyFunc.h` is in a directory called `MyDir` on disk A. Note that this form of the include statement is slightly different from that used previously for header files of the standard libraries, for example,

$$\text{\#include <math.h>} \tag{14.6}$$

In (14.6), the angle brackets < and > indicate that the file will be in one of the standard directories of the system, and we do not need to specify a path. For our own header file, however, the path as given in (14.5) is enclosed in quotes rather than braces. Alternatively, we could use the form

```
#include "MyFunc.h"
```

without the full path and the compiler would look in the user's current directory for this file and then in other "standard" files.

Implementation File

Let us assume that our function programs have been put into another file named `MyFunc.cpp`, which is called the *implementation* or *source file*. (Note the different file extension, which distinguishes this file from the header file `MyFunc.h`.) Then, when we compile and run `main()`, programs from `MyFunc.cpp` are linked with `main()` so that the actual function programs are available when `main()` needs them. In this case, Figure 14.1 changes to the more elaborate schematic Figure 14.2.

As indicated in Figure 14.2, the header and source files will first go through a preprocessor that will send source files ready for compilation to the compiler. The

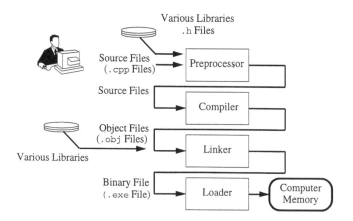

Figure 14.2 Another Schematic of Compiling and Linking

compiler in turn produces what is called an *object* file, which will usually be stored on your disk with an `.obj` file extension (in Windows). It is this object file that is sent to the linker, which will combine it with other programs that have been precompiled. The linker sends binary machine language programs to the loader, and will usually also store them on your disk with an `.exe` extension (again, in Windows). Finally, the loader decides where in main memory your programs and data will be put.

If the directive (14.5) is put before `main()` and the file `MyFunc.h` includes all the prototypes of functions used in `main()`, then `main()` can be compiled without any further knowledge of the functions being used. One advantage of this is that every time we need to compile `main()` to make corrections or changes, we do not need to compile the functions it will use. We only need to recompile the functions if we make changes in them. Of course, when the program is run, it will need the function programs.

Also, as indicated in Figure 14.2, there is the possibility of linking with "various" libraries. One of these may be your own personal library (or libraries; you may

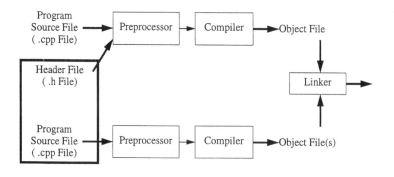

Figure 14.3 Combining Programs from Different Compilations

establish as many as you wish), as discussed above. In particular, if one of your libraries had already been compiled, programs from it would be linked to the rest of the program at this time. But you may also link with still other libraries, perhaps ones containing functions developed by friends or colleagues, or commercial libraries. The process of combining different programs from different files is indicated in the schematic of Figure 14.3. The specific details as to how you do the linking process will depend on the particular compiler and operating system you are using.

MAIN POINTS OF CHAPTER 14

- Actual parameters in a function may be called by value, in which case the value of the variable is passed to the function, or called by reference, in which case an address is passed to the function. Arrays are always passed by reference, whereas fundamental types are passed by value unless the reference declarator, &, is used. The return value of a function may also be a reference.

- Functions may be recursive; that is, they may call themselves.

- Functions may be declared to be inline, in which case the compiler will try to write the code for the function directly where it is needed.

- Function names may be overloaded so that the same name is used for different functions. This should only be done for functions that do essentially the same thing, but with different numbers and/or types of arguments.

- A function may have default parameters that may be omitted when the function is called. Such parameters must be given default values in the function definition and put after any nondefault parameters.

- Local variables within a function may be declared to be static, in which case the variable retains its value between function calls.

- Templates offer a way to generate automatically functions differing only in the types of the arguments and return values.

- Personal libraries of functions may be defined. Prototypes of the functions should be included in a header file, and the function programs themselves in a separate implementation file. This implementation file may be compiled independently of `main()` and functions then linked to `main()` at run time.

EXERCISES

14.1 Suppose that a is an array of floating point numbers with the values:

    a[0] = 1.0, a[1] = 2.0, a[2] = 4.0, a[3] = 6.0

If the function of Example 14.1 is called by `p_prime(a, d, 3)`, what are the values of `d`? Are the parameters in this function called by value or reference?

14.2 Write the prototype for a function that will call by reference an `int` variable `i` and a `float` variable `x`.

14.3   Define a function by

```
int f(int i, int &j){
  j = i + i;
  return i * i;
}
```

What are the values of k and m in `main()` after the statements

```
int m = 1;
int k = f(2, m);
```

14.4   The *Fibonacci numbers* are defined by the recurrence

$$f_{n+1} = f_n + f_{n-1}, n = 1, 2, \ldots, f_0 = 0, f_1 = 1$$

Write a recursive function with n as a parameter and return f_n as the function value. Also, write a nonrecursive function to compute f_n and comment on the difference between the two functions.

14.5   Declare the function for the integrand in Example 6.2 to be inline and place it in the proper position. Then discuss what happens when this function is needed in Example 6.1.

14.6   The cosine of the angle between two vectors (x_1, x_2) and (y_1, y_2) is

$$\cos \theta = \frac{x_1 y_1 + x_2 y_2}{\text{length}(x_1, x_2)\text{length}(y_1, y_2)}$$

where `length()` is given by equation (14.3a). Likewise, in 3-space the cosine between the vectors (x_1, x_2, x_3) and (y_1, y_2, y_3) is, with `length()` given by (14.3b),

$$\cos \theta = \frac{x_1 y_1 + x_2 y_2 + x_3 y_3}{\text{length}(x_1, x_2, x_3)\text{length}(y_1, y_2, y_3)}$$

Write a function to compute the cosine for vectors in 2-space and then overload the function name for the function to compute the cosine for vectors in 3-space. Write a `main()` program that calls both functions.

14.7   Redo Exercise 9.10 as follows. Write a single function that computes $\bar{x}$ and σ by making them both reference parameters.

14.8   Indicate which of the following are legal or illegal statements and explain your answer.

a.   `float g(float a, int k = 10); //function prototype`

b.   `int f(int a, int b); //two function prototypes`
`    int f(int a, int b, int c); //in same program`

14.9 Write a function, as discussed in Section 14.3 under Static Variables, that has a local variable that counts the number of times the function has been called.

14.10 Write two functions and put them in an implementation file separately from `main()`. Build a header file that contains prototypes for your two functions and have an include statement for this header file before `main()`. Now compile the implementation file for the functions and then compile `main()` separately. Finally, run `main()` and verify that the function programs are linked to `main()`. Note: You may need additional information about your compiler to do this exercise.

14.11 Write a function whose prototype is

```
float& maximum(float a[ ], int n)
```

that will find the maximum value `a[m]` in an array `a` of length `n` and return the address of `a[m]`. Then use this function in a function `main()` that includes the statement

```
maximum(a, 5) = 0;
```

Now remove the reference declarator `&` from the function definition and note the effect on your program.

14.12 Use either a debugger or temporary print statements (see Section 2.5) to step through the recursive function of Example 14.3 and watch the successive calls of the function.

14.13 Use the template of Example 14.5 to automatically generate `max` functions for `float` and `int` numbers, as discussed in the text. Test your program to verify that these functions are generated correctly.

chapter
15
POINTERS

In this chapter, we introduce pointers and their basic properties. An array name is an important type of pointer, and we pursue this connection in Section 15.2. In Section 15.3, we show how to use pointers to write a matrix-vector multiplication function that is valid for matrices of any size, in contrast to the function in Example 12.2. In Section 15.4, we use pointers in conjunction with character strings and introduce some useful string functions.

15.1 POINTER VARIABLES

Consider the following statements:

$$\text{int i = 5;} \qquad (15.1a)$$

$$\text{int *ptr;} \qquad (15.1b)$$

$$\text{ptr = \&i;} \qquad (15.1c)$$

The second statement, (15.1b), defines `ptr` to be a *pointer variable*; it will take on values that are memory addresses of integer variables. The third statement, (15.1c), assigns such an address to `ptr`; here, `&` is the *address of* operator and is technically different from the reference declarator used in the previous chapter. The reference declarator is used with function parameters, whereas the address of operator is used as in (15.1c) to assign to `ptr` the address of the integer variable i. [Note that `&` cannot be used on constants or expressions; for example, `&10` or `&(i + j)` are not legal.] We could, if we wished, combine (15.1b) and (15.1c) to initialize `ptr` as it is declared:

```
int *ptr = &i; //initialization
```

Figure 15.1 is a memory diagram showing the relationship of `ptr` and i. In this

Figure 15.1 Pointers and Memory

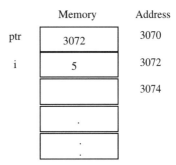

diagram, i is stored in memory location 3072. The value of ptr is this address and is stored in another memory location, with address 3070 in the diagram.

It is the asterisk in (15.1b) that determines that ptr is a pointer variable rather than an ordinary integer variable. The asterisk could also be attached to int rather than ptr, as in int* ptr. However, (15.1b) is the form that extends to multiple *Declaring Multiple* variables, as in
Pointers

```
int *ptr_1, *ptr_2; //ptr_1 and ptr_2 are pointers
```

If we wrote instead

```
int* ptr_1, ptr_2; //ptr_2 is not a pointer
```

then ptr_1 would be a pointer, but ptr_2 would be an ordinary integer variable. We say that * *binds* to ptr_1 and not to int.

Pointer variables may also be declared for any other fundamental or user-defined data type, as in the following declarations:

$$\texttt{float *fptr;} \tag{15.2a}$$

$$\texttt{char *cptr;} \tag{15.2b}$$

$$\texttt{double *dptr;} \tag{15.2c}$$

$$\texttt{car *car\_ptr;} \tag{15.2d}$$

In (15.2d), car is the structure defined in Example 11.3. In both (15.1) and (15.2) we have used ptr or a variation as the name of the pointer; this is not necessary, and any identifier may be used as a pointer name.

Alias   Next, suppose that i and ptr are as in (15.1). Then the statement

$$\texttt{*ptr = 10; // equivalent to i = 10} \tag{15.3}$$

is equivalent to i = 10. We say that ptr is an *alias* for i because after the assignment ptr = &i, both i and *ptr refer to the same variable. The * in (15.3) is called the *indirection* or *dereferencing* operator: *ptr is the contents of the memory location
Dereference or whose address is ptr. (We note that void *p is a valid declaration; it declares a
Indirection pointer to an object of unknown type, but such a pointer can not dereferenced.) *ptr

can not only be used in an assignment statement like (15.3) but can also be used any place the variable `i` would be, as in the expressions

$$*ptr * 2 \qquad //same\ as\ i * 2 \qquad\qquad (15.4a)$$

$$*ptr + *ptr + j \quad //same\ as\ i + i + j \qquad (15.4b)$$

$$(*ptr)++ \qquad //same\ as\ i++ \qquad\qquad (15.4c)$$

In each instance of (15.4), it is the current value of `i` that is used in the expression. Note that, as in (15.4c), the increment operator `++` (or the decrement operator `--`) can also be used, just as with `i` itself.

Pointers to Pointers   We may also define *pointers to pointers* by, for example,

$$int \ \ **pptr; \ //pptr\ is\ pointer\ to\ int\ pointers \qquad (15.5)$$

Then the statement

$$pptr = \&ptr; \ //pptr\ is\ address\ of\ ptr$$

assigns to `pptr` the address of the pointer `ptr`. If `ptr` has been defined by the statements (15.1), then the statement

$$**pptr = 10; \ //equivalent\ to\ i = 10$$

assigns the value 10 to `i`. Here, `*pptr` dereferences `pptr` to give `ptr`, so that the second dereferencing is the same as (15.3).

Finally, we may make pointer comparisons using the relational operators `==` and `!=` (and also `<`, `>`, `=>`, `<=` since the value of a pointer is an address, which is just an integer). For example, if `ptr_1` and `ptr_2` are both pointers to integers, we could write

$$if(ptr\_1 == ptr\_2) \ cout << *ptr\_1;$$

or

$$if(ptr\_1 \ != ptr\_2) \ cout << *ptr\_2;$$

To make such comparisons, both pointers should point to variables of the same type; thus, if `ptr_i` pointed to integers and `ptr_f` pointed to floating point numbers, the expression

$$ptr\_i == ptr\_f$$

would at least generate a warning.

Constant Pointers[†]

Constant Pointer   The statement (15.1b) defines `ptr` to be a pointer variable that may take on different values. The statement (15.1c) assigns to `ptr` a particular value, the address of `i`. If `count` were another variable of type `int`, the statement

---

[†]This subsection may be omitted on first reading.

$$ptr = \&count; \qquad\qquad (15.6)$$

would assign a new value, the address of count, to ptr. In contrast to pointer variables, there are also *constant pointers*. The statement

$$int \quad const \quad *ptr = \&j; \qquad\qquad (15.7)$$

defines ptr to be a constant pointer whose value is the address of j. As a constant pointer, ptr cannot be changed; for example, the statement (15.6) cannot now be used to assign a new value to ptr. The value of the variable j in (15.7) can, of course, be changed. On the other hand, the statements

Pointer to Constant

```
const int j = 10;
const int *ptr = &j;
```

now define ptr to be a pointer variable that can point to constants. Thus, here, ptr can be changed by a statement such as

$$ptr = \&k;$$

where k is another constant. Even though the pointer ptr may be changed, the constant values for j and k cannot, of course, be changed. Finally, we may have *constant pointers to constants* such as defined by

Constant Pointer to Constant

$$const \ int \ j = 10; \qquad\qquad (15.8a)$$

$$const \ int* \ const \ ptr = \&j; \qquad\qquad (15.8b)$$

Here, both j and ptr are declared to be constants; so neither may be changed.

15.2 POINTERS AND ARRAYS

Array Name Is a Pointer

Another, and very important, type of pointer is an array name. (An array name is actually a constant pointer, as defined in the previous subsection.) If an array a has been defined by

$$float \quad a[10] \qquad\qquad (15.9)$$

we may write

$$r = *a; \quad //same \ as \ r = a[0]$$

to access the first element of a. This is valid because the array name, a, is just the address where the array begins so that *a is the contents of that address. We may also access any other element of a by the construction

Pointer Arithmetic

$$r = *(a+i); \quad //same \ as \quad r = a[i] \qquad\qquad (15.10)$$

The statement (15.10) illustrates *pointer arithmetic*. a is the initial address of the array, and if a is declared by (15.9), then each element of a requires 4 bytes of storage. Hence, $a + i$ must be the address of the $(i + 1)$st element of the array and $4 * i$ must be added to a to obtain this address. This is done automatically by the compiler: The number i in (15.10), which is called the *offset*, is *scaled* by the size of the type so as to give the correct address for the type of the array. The memory diagram of this situation is shown in Figure 15.2, in which the addresses differ by four for floating point numbers.

Pointers to Arrays   We may also define pointers to arrays, as in the following example:

```
float a[10], *ptr;
//assign values to a here
ptr = &a[0]; // or we could write ptr = a
float sum = 0;
for(int i = 0; i < 10; i++)
    sum = sum + * (ptr++); //can write as sum += *ptr++
```

This code sums the elements of the array a. Note that we may set the value of ptr to be the beginning address of a by either ptr = &a[0] or ptr = a. The key statement in the above code fragment is the last one. When i = 0, *ptr is the value of a[0] and this is put in sum; then ptr is incremented so that for i = 1, *ptr is the value of a[1], and so on. Note that * (ptr++) can be written as *ptr++ and is different from (*ptr)++, as used in (15.4c).

Another way to write the last statement in the above example is

$$sum = sum + * (ptr + i)$$

Pointer Is an Array   since ptr + i points to a[i]. Comparing this with (15.10), it is tempting to write
Name Alias

$$sum = sum + ptr[i]$$

and this is permissible. Thus, once we have set ptr to the first address of the array a, we can treat ptr as an alias for a and use array notation with ptr as well as with a.

As another illustration of the correspondence between pointers and arrays, consider the statements:

$$float \quad a[10];$$

(15.11)

$$float \quad *ptr\_a = a - 1;$$

This defines ptr_a to be a pointer initialized to the beginning address minus 1 of the array a. Thus ptr_a[1] is the address of a[0] and, in general, ptr_a[i] is the address of a[i - 1]. Thus we may view ptr_a as an array whose indices start at 1, not 0. This is one way to allow array indices to begin at 1 rather than 0 in order to keep the usual mathematical notation for vectors.

Figure 15.2 Memory for an Array

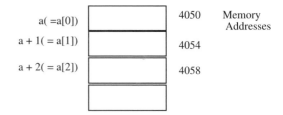

Arrays of Pointers

As well as pointers to arrays, we may also define arrays of pointers. For example,

```
float *ptr[4];
```

defines an array of 4 pointers to floating point numbers. If a and b are two arrays of floating point numbers defined by

```
float a[20], b[4];
```

the statements

```
ptr[0] = &a[1];
ptr[1] = &a[8];
ptr[2] = &a[15];
ptr[3] = &a[19];
```

assign to the elements of ptr the addresses of the indicated elements of the array a. Then the loop

```
for(int i = 0; i < 4; i++)
    b[i] = *ptr[i];
```

assigns to the array b the elements 1, 8, 15, and 19 of the array a.

15.3 POINTERS AND FUNCTIONS

Consider the function in Example 15.1. This function has one argument, which is a pointer to variables of type int. When the function is called, its argument must be a pointer; for example, a possible call is

$$func(\&j) \qquad\qquad (15.12)$$

where j is a variable of type int, and thus &j is a pointer to j. In arguments of functions, it is usual to place the * as shown in Example 15.1, rather than as in int *i, but either form is acceptable.

EXAMPLE 15.1
A Function with a Pointer Argument

```
void func(int* i){
    *i = *i + *i;
}
```

When the function of Example 15.1 is called by (15.12), the address of j is the argument so that the statement in the body of the function is equivalent to

$$j = j + j;$$

This is another case of call by reference, as discussed in the previous chapter.

EXAMPLE 15.2
A Function with an Array Argument

```
void f(float c[ ], int n){
   for(int i = 0; i< n; i++)
      c[i] = i;
}
```

As another example, consider the function in Example 15.2. If in `main()`, an array is defined by

$$float \quad a[10]; \tag{15.13}$$

then the function of Example 15.2 could be called by, for example,

$$f(a, 6) \tag{15.14}$$

and the function would set the first six elements of a to $0, 1, \ldots, 5$. As discussed in Chapters 9 and 14, it is the beginning address of the array a that is passed to the function. Note that we would *not* write `&a` for this address since the array name itself is a pointer, that is, an address. We could call the function by `f(&a[0], 6)`, although this is clumsy compared with `f(a, 6)`. However, the address of operator is useful if we wish to pass other than the beginning address of the array, as in `f(a[2],4)`. In this case, we could also use pointer arithmetic to write `f(a + 2, 4)`.

Pointers Can Replace Array Names

The function of Example 15.2 may be written in a different form, as shown in Example 15.3. If this function is again called by (15.14), where a is declared by (15.13), then the argument a acts as a pointer to the array and the function again sets the first 6 elements of the array to $0, 1, \ldots, 5$. We could replace the statement `ptr[i] = i;` by `*ptr++ = i;` in Example 15.3, and this statement behaves as discussed previously: When the function is called by (15.14), the address of a is passed to the function; thus `*ptr` is equivalent to `a[0]`, so that the first time through the `for` loop of Example 15.3, `a[0]` is set to 0. `++` then increments `ptr` to obtain the address of `a[1]`, and so on. Although the functions of Examples 15.2 and 15.3 seem to be quite different, they are actually the same function and will be compiled in the same way.

EXAMPLE 15.3
The Same Function with a Pointer Argument

```
void f(float* ptr, int n){
   for(int i = 0; i < n; i++)
      ptr[i] = i; //could also write *ptr++ = i;
}
```

Two-Dimensional Arrays

We next give an example of the use of pointers with two-dimensional arrays. Recall that in Chapter 12 we wrote a function (Example 12.2) for computing the product of a matrix and a vector. That function had a serious deficiency: the second dimension of the input array A representing the matrix had to be given explicitly, and this precluded using the function except for matrices of that specific size (2×2 matrices in Example 12.2). By means of pointers we may overcome this difficulty, as illustrated by the program in Example 15.4.

The first three arguments of the matvec() function in Example 15.4 are pointers to the arrays A, x, and b. Note that we have used the array names themselves as the pointer names, although other names such as ptr_A could have been used. We could also have used the usual array notation x[ ] and b[ ] for these two parameters in the function heading, but in this example we must use float* A for A.

EXAMPLE 15.4
Matrix-Vector Multiplication Using Pointers

```
#include <iostream.h>
void matvec(float* A, float* x, float* b, int n);
int main(){
  float A[2][2] = {{3.,6.},{9.,7.}};
  float b[2],x[2]={4.,6.};
  int n = 2;
  matvec(A[0], x, b, n);
  cout << "b[0]=" << b[0] << " b[1]=" << b[1];
  return 0;
}
void matvec(float* A, float* x, float* b, int n){
  for(int i = 0; i < n; i++){
    float temp = 0;
    float *rowptr = A + i * n; //address of row i
    for(int j = 0; j < n; j++)
      temp = temp + rowptr[j] * x[j];
    b[i] = temp;
  }
}
```

In the definition of the function matvec() the statement to change temp, as well as the following statement, use the usual array notation for x[j] and b[i]. However, for the two-dimensional array A, we cannot use the array notation A[i][j]. Instead, we view the two-dimensional array as a long one-dimensional array consisting of the first row, followed by the second row, and so on, as discussed in Chapter 12.

Pointer Arithmetic Then we do pointer arithmetic to access array elements. Thus

$$A + i * n + j \tag{15.15}$$

is the address of A[i][j]. This is illustrated in Figure 15.3, in which the beginning element in row 0 has address A, the beginning element in row 1 has address A + n,

```
row 0, address A          A[0][0] ... A[0][j] ... A[0][n-1]
row 1, address A + n      A[1][0] ... A[1][j] ... A[1][n-1]
  .
  .
  .
row i, address A + i * n  A[i][0] ... A[i][j] ... A[i][n-1]
```

Figure 15.3 Schematic of the Array A

and so on until the beginning element in row i has the address A + i * n. Thus the address of A[i][j] is as in (15.15). As indicated in Figure 15.3, A + i * n may be considered to be a pointer to row i of the array A. The function matvec() in Example 15.4 is written by defining the pointer rowptr with this value rather than using the expression (15.15) in the statement to change temp; with some compilers this will be more efficient since it ensures that the address A + i * n is evaluated only once for each i.

Calling the Function

When we call the function matvec() in main() in Example 15.4, the array names x and b serve as pointers to those arrays. With two-dimensional arrays, however, the situation is different. In the function call the array name A would be considered to be a pointer to a one-dimensional array of arrays (as discussed in Chapter 12, a two-dimensional array may be considered to be a one-dimensional array of arrays), but the function definition requires the argument to be a pointer to floating point numbers. Thus there is an incompatability. This is resolved by using the argument A[0] in the function call; A[0] is a pointer to the first row of A and hence acts in the same way as the array names x and b. Alternatively, we could use (float*)A in place of A[0]; this casts A so it is treated as a pointer to floating point numbers. Still another option is &A[0][0], where & is the address of operator. (See Exercise 15.6.)

In the next chapter, we will give another approach to a matrix-vector multiplication function using pointers to pointers. This will allow us to use the usual array notation A[i][j] within the function. Although the use of the matvec() function in Example 15.4 is illustrated only for a 2 × 2 matrix A, the function can handle general $n \times n$ matrices (see Exercise 15.5).

Pointers to Functions

In addition to pointing to fundamental types, structures, and arrays, we may also define *pointers to functions*. For example, the statement

```
int (*fptr)(int);
```

defines a pointer fptr to functions that have a single int argument and have a int return value. (The parentheses about *fptr are necessary here.)

As an example of how we might use pointers to functions, suppose that we have the following statements that form the sum of several values of an integer-valued function f:

```
        int p = 1;
        for(int i = 1; i <= n; i++)
          p += f(i); //same as p = p + f(i)
```

Now suppose that we wished to do this for a number of different functions, and different values of n. Writing this code over and over for different functions would be very tedious. What we would like is a function that accepts a function as an argument. We can do this by means of pointers to functions, as illustrated in Example 15.5.

In Example 15.5, we define a function sum() whose arguments are an integer n and a pointer to integer-valued functions with a single integer argument. In main(), we call this function twice with different functions f1 and f2. (Prototypes for f1 and f2 are given but the function definitions themselves are not shown. See Exercise 15.8.) In the first call of sum(), it is the function f1 that is utilized, whereas in the second call it is f2.

EXAMPLE 15.5
Using Pointers to Functions

```
        #include <iostream.h>
        int f1(int); //prototype
        int f2(int); //prototype
        int sum(int(*f)(int), int); //prototype
        int main(){
          int n1, n2;
          cout << "enter n1 and n2" << endl;
          cin >> n1 >> n2;
          cout << sum(f1, n1) << endl;
          cout << sum(f2, n2) << endl;
          return 0;
        }
        int sum(int(*f)(int j), int n){
          int value = 0;
          for(int i = 1; i <= n; i++)
            value += f(i);
          return value;
        }
```

Being able to use pointers to functions in this way is a very powerful feature, and is potentially useful in a number of situations. For example, the integration program of Examples 6.1 and 6.2 is restricted to a particular integrand function f. If the integrand changes, the program must be run again for the new function. But if we modified the heading of the rectangle function of Example 6.1 to

```
        float rectangle(float a, float b, int n, float(*f)(float x))
```

then we could call this function for several different integrands. Exercise 15.9 asks you to make this modification.

15.4 POINTERS AND STRINGS

Recall from Chapter 11 that a string of characters may be defined as an array of characters. For example, the declaration

$$\text{char string\_one[17] = "This is a string";} \qquad (15.16)$$

defines a string of 17 characters (including the termination character) as elements of the array `string_one`. Alternatively, we may define a string by means of a pointer, as in

$$\text{char *string\_two = "This is also a string";} \qquad (15.17)$$

Note that `string_two` is a pointer variable [initialized as in (15.17)], whereas `string_one`, as an array name, is a constant pointer.

String Functions

We next return to the program of Example 12.3, in which we searched each string for occurrence of the word "dog" and then replaced "dog" with "cat." In Example 12.3, we tested each character separately, and likewise replaced each character separately. It would be much better to be able to compare and replace whole strings, rather than just individual characters, and we can do this by means of *string functions*, contained in the library `string.h`. This library has numerous functions for string manipulations, but we will be concerned here with only the following four functions:

$$\text{strlen(char *ptr\_s)} \qquad (15.18)$$

This function returns the number of characters before the termination character in the string pointed to by `ptr_s`.

$$\text{strncmp(char *ptr\_s1, char *ptr\_s2, int n)} \qquad (15.19)$$

This function compares up to n characters (characters that follow the termination character are not compared) from the string pointed to by `ptr_s1` to the string pointed to by `ptr_2`. It returns the value 0 if these strings are equal.

$$\text{strncpy(char *ptr\_s1, char *ptr\_s2, int n)} \qquad (15.20)$$

This function copies up to n characters from the string pointed to by `ptr_s2` into the string pointed to by `ptr_s1`, assumed to have at least n characters. A related function does not use the `int` argument and copies the whole string:

$$\text{strcpy(char *ptr\_s1, char *ptr\_s2)} \qquad (15.21)$$

We now give in Example 15.6 a program that utilizes these functions and generalizes the replace "dog" by "cat" program of Example 12.3. In this program, we read in a test string and a replacement string. Then in each word `s[i]` of the array, whenever the test string is matched, it is replaced by the replacement string. We assume that the test string and the replacement string are the same length, and we determine that length by the `strlen` function. (Exercise 15.10 asks you to add a test to the program to check that the replacement string is indeed the same length as the test string.)

EXAMPLE 15.6
Replacing Strings

```
//This program searches a "sentence" of up to 100 words, each
//word up to 20 characters long, and replaces a test string by
//another string assumed to be of the same length.
#include <iostream.h>
#include <string.h> //to use string functions
int main(){
  char s[100][21],test[21],replace[21];
  int n, len_s;
  cout << "What is the test string?\n";
  cin >> test;
  int len = strlen(test);
  cout << "len =" << len << endl;
  cout << "what is the replacement string?\n";
  cin >> replace; //assumed same length as test string
  cout << "how many words?\n";
  cin >> n;
  for(int i = 0; i < n; i++){
    cout << "read a word\n";
    cin >> s[i];
    len_s = strlen(s[i]); //length of ith word
    cout << "len_s=  " << len_s << endl;
    cout << "input word is " << s[i] << endl;
    for(int j = 0; j <= len_s - len; j++)
        if(strncmp(test, &s[i][j], len) == 0) //search for string
            strncpy(&s[i][j],replace,len); //replace string
    cout << "new string is " << s[i]    << endl;
  }
  return 0;
}
```

*Description of
Example 15.6*

The first two arguments of both the strncmp and strncpy functions must be pointers to strings, and the array names test and replace are pointers to those strings. For the two-dimensional array s, however, we must use the address of operator, &, to obtain a suitable pointer. Thus &s[i][j] points to character j in string i, and the j for loop successively looks at each substring of length len in s[i] to see if it matches the test string test; if so, it replaces that string with the string in replace. Note that we use the strlen function to obtain the length, len_s, of the string s[i] and use this length to determine how large j should become: len_s - len. For example, if len = 3 and len_s = 6, s[i][3] starts the last string in s[i] to be tested.

MAIN POINTS OF CHAPTER 15

- Pointer variables take on addresses as their values, and may be defined to point to any data type, including user-defined types. A declaration of a pointer variable is of the form int *ptr.

- * is the dereferencing or indirection operator: `*ptr` is the contents of the memory location whose address is `ptr`. `&` is the address of operator: `&i` is the address of the variable `i`.

- Pointers may be constant as well as variable. An array name is an example of a constant pointer.

- Pointer arithmetic may be performed, and the compiler automatically scales to give the correct address for the data type.

- Arrays of pointers may be declared, and pointers may be used as function arguments and returned by functions.

- Pointers are useful in working with two-dimensional arrays, in particular, in defining functions of two-dimensional arrays whose size may vary. A matrix-vector multiplication function is an example; two-dimensional arrays of characters (arrays of strings) are another example.

- `string.h` is a library of string manipulation functions. Three useful functions in this library are `strlen`, to obtain the length of a string, `strncomp`, to compare two strings, and `strncpy`, to copy one string into another.

EXERCISES

15.1 Declare pointers to variables of type `int` and `float`. Then dereference these pointers in assignment statements.

15.2 What is the value of the variable `i` after the following statements?

```
int i = 4;
int *ptr = &i;
i = *ptr + *ptr;
```

15.3 What is the value of `a[2]` after the following statements?

```
float a[3] = {1.1, 2.2, 3.3};
a[2] = *a + *(a + 1) + *(a + 2);
```

15.4 Define a function, along the lines of Example 15.4, that has two pointers and an `int` as arguments and computes the sum of two arrays.

15.5 Change the function `main()` of Example 15.4 to do matrix-vector multiplication for 3 × 3 and 4 × 4 matrices. Run the new programs.

15.6 Replace `A[0]` in the function call of `matvec()` in Example 15.4 by `(float*)A` and `&A[0][0]`, as discussed in the text, and verify that these both work also.

15.7 Using pointers, modify the programs of Exercises 12.3 and 12.4 so that they can handle matrices of any sizes.

15.8 Define two functions `f1` and `f2` and run the program of Example 15.5.

15.9 Modify the function `rectangle` of Example 6.1 so that one of its parameters is a pointer to functions, as discussed in the text. Then modify the function `main()` of Example 6.2 so that it calls `rectangle()` for at least two different integrand functions.

15.10 Run the program of Example 15.6 for various "sentences" and `test` and `replace` strings. Then add a test to the program that determines whether the replacement string is indeed the same length as the test string. Finally, modify the program so that the replacement string does not need to be the same length as the test string.

15.11 Write a function that will count the number of characters in a string, not including the termination character. The function heading should be the same as that of `strlen()`.

16

DYNAMIC MEMORY

It would be nice to be able to define arrays as in

```
float x[m], b[n]; //illegal if m and n not constants
```

where n and m are integer variables. As discussed previously, this is only possible if n and m have been previously declared as const int and numerical values assigned to them. However, we can achieve *dynamic memory* allocation in a different way by means of pointers and the new operator. In fact, in C++ probably the most important use of pointers in scientific computing is in conjunction with dynamic memory allocation of arrays.

Dynamic memory is called dynamic because the amount of memory used and the duration of its use is under program control at run time. This is in contrast to fixed-size and fixed-duration allocation of memory, which is done when the program is compiled. For example, consider the code fragment shown in Example 16.1:

EXAMPLE 16.1
Static Allocation of Memory

```
const int size = 10,000;
float a[size], b[size], c[size];
void assign() {
  float d[size];
  //more code here
};
int main()
//more code here
```

In Example 16.1 the compiler allocates space for three global arrays in the global variable portion of memory. This space is reserved for the entire duration of the program. Memory for the array d is allocated on what is called the *stack* on entry to the function `assign()`. The memory remains allocated as long as the function is executing, but ceases to exist on exit from the function. The most important aspect of both of these allocations is that the amount of memory allocated is fixed and determined when the program is compiled. Thus we must know the exact amount of memory needed, or at least an upper bound on the amount needed.

16.1 DYNAMIC MEMORY ALLOCATION

Dynamic memory is quite different. The important characteristics of dynamic memory are:

1. Any amount of memory may be requested (up to the amount of memory available for the entire program);

2. The memory remains allocated until it is explicitly deleted.

Thus we no longer need to know the size of arrays when the program is compiled. Instead, we can allocate space as it is needed.

Dynamic memory is allocated using the `new` operator:

```
float *b = new float[n]; //assume int n has a positive value
// Returns a pointer to 4 * n bytes of contiguous memory.
```

The new *Operator*

The amount of space requested is specified by the integer n, which may be either a constant or a variable (which has been given a value). The operator `new` initializes the pointer b to the address of the correct amount of space if the space is available, or a 0 if there is insufficient space to honor the request. (In a new C++ standard, an "exception" will be issued if insufficient space exists.) Thus in the above example b points to 4 * n bytes of contiguous storage to hold n floating point numbers. A schematic of memory illustrating this is shown in Figure 16.1.

The Free Store

The memory assigned by the `new` operator comes from a pool of memory called the *free store*. One can think of the free store as a large bag full of wooden blocks (memory). Whenever you need some memory, you reach into the bag using the `new` operator and pull out the number of blocks needed. Just like a real bag of blocks there is only a finite number. If you continue to take blocks from the bag and never return any, then eventually the bag will become empty, and the `new` operator will return a 0. To avoid this situation, memory should be deallocated when it is no longer needed, and this is done by using the `delete` operator:

Figure 16.1 Allocating Dynamic Memory

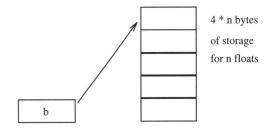

4 * n bytes of storage for n floats

b

```
                         delete [ ]b;
```

Example 16.2 illustrates the above statements in a complete program that allocates storage for two arrays.

EXAMPLE 16.2
Assigning Memory with the new Operator

```
#include <iostream.h>
int main(){
  cout << "what is n?\n";
  int i, n;
  cin >> n;
  float *a = new float[n];
  float *b = new float[n];
  cout << "what is a?\n";
  for(i = 0; i < n; i++)
    cin >> a[i];
  for(i = 0; i < n; i++){
    b[i] = a[i];
    cout << "b[" << i << "] is " << b[i] << "\n";
  }
  delete [ ]a;
  delete [ ]b;
  return 0;
}
```

In the program of Example 16.2, a and b are defined to be pointers, and the new operator assigns storage based on the input parameter n. a and b may now be treated as *n*-long arrays, as evidenced by the remaining statements. Note that we cannot initialize arrays when the new operator is used to assign them storage. (This restriction does not apply to class objects, to be discussed in the next chapter.)

At the end of the program, we use the delete operator, which removes the storage assigned to a and b. Returning to the bag of blocks analogy, the delete operator returns the blocks to the bag for subsequent reuse. The use of [ ] in these delete statements is necessary since storage has been assigned for an array. In general, whenever [ ] is used with the new operator, it should also be used with the delete operator.

Deleting Storage

Strictly speaking, these delete statements are not needed in Example 16.2 since upon program termination the storage is automatically released. However, it is very good practice to release memory when you are through with it. This is especially important when using the new operator within functions since when the function terminates, the pointer will be destroyed, but the memory will remain allocated and unusable. We note that it is not possible to delete memory not allocated by the new operator.

Persistence

We next illustrate a second major difference between statically and dynamically allocated memory: *persistence*. Example 16.3 illustrates a rather common error. The intent of the function in Example 16.3 is to return a pointer to an array that has been assigned values within the function. `float*` in the function heading indicates that the return type is a pointer to floating point numbers, and that is indeed what b is, since it is an array name. The problem is that the memory allocated to the array b *ceases to exist upon exiting the function*, so b now points to memory that may be overwritten in future statements in the program. Thus the memory assignment to b does not *persist* beyond the function. But we can solve this problem easily by using dynamic memory allocation, as shown in Example 16.4.

A Common Error

EXAMPLE 16.3
Bad: Return of Pointer to Local Variable

```
float* assign(int n){          //This
  float b[20];                 //Function
  for(int i = 0; i < n; i++)   //is in
    b[i] = i;                  //error
  return b;
}
```

EXAMPLE 16.4
OK: Return of Pointer to Dynamic Memory

```
float* assign(int n){
  float *b = new float[n];     //This
  for(int i = 0; i < n; i++)   //Function
    b[i] = i;                  //is
  return b;                    //ok
}
```

Dynamic memory continues to exist, independent of scope, until it is deleted by a `delete` operation. Thus in the function of Example 16.4, the memory assigned to b is not deleted upon exiting the function since dynamic memory is only released by the `delete` operation. Hence, outside the function, the function value indeed points to the memory that was assigned to b within the function. With this version of the function we allocate exactly enough memory for the return result, and this memory will remain allocated until explicitly deleted.

Testing for Insufficient Memory

A potential problem with dynamic memory allocation is that the `new` operator may not be able to allocate sufficient memory. This is not likely to happen in small

Null Pointer

programs, but in large programs that use lots of memory it could be a problem. As mentioned earlier, if the new operator is unable to allocate storage, instead of a pointer to storage it returns 0, which represents the *null pointer*. To be prudent, we can test to see if the null pointer has been returned. For example, after the statement in Example 16.2 that allocates storage for a, we could have the statement

$$\text{if(a == 0) cout << "Storage could not be allocated"; \qquad (16.1)}$$

You may also see 0 represented in this context by `'\0'`, which is the character whose binary representation is all 0s (and as used as the termination character for character arrays discussed in Chapter 11). You may also see a statement such as (16.1) written in the form

```
if(a == NULL) cout << "Storage could not be allocated";
```

Here, NULL stands for the null pointer, and may be used in place of 0.

Common Mistakes with Dynamic Memory

We end this section with a discussion of two common mistakes when using dynamic memory in C++: dangling references and garbage. (C++ does not provide automatic "garbage collection" as do some other languages.) Dangling references can lead to strange and difficult errors to correct, while an excess of "garbage" wastes memory and may prevent the program from completing its work.

Dangling Reference

A *dangling reference*, also called a *dangling pointer*, occurs when a pointer points to memory that has been deallocated. This can occur in two ways. First, the memory is deleted but the pointer variable not reset to 0. For example, the code fragment

```
float *ptr = new float [some_value];
// some code that uses ptr
delete [ ] ptr;
```

produces a dangling reference. In and of itself this is not a problem. The problem arises if ptr is used again without first allocating new memory. Then one of two equally bad things may happen. First, if the memory that ptr points to has been reallocated, any assignments to ptr, such as ptr[1]=4.5;, will corrupt the new data, and this may not be observed until much later during program execution, making this sort of error very difficult to detect. On the other hand, if the memory has not been reallocated, everything will be fine *until* the memory is reallocated to other data, at which point anything previously assigned using ptr will be lost.

A good way to prevent, or at least detect, this sort of error is always to set pointers to zero once the memory has been deallocated:

```
delete [ ] ptr;
ptr = 0;
```

By setting ptr to zero we cause the program to fail if ptr is dereferenced after the deletion, for example, by ptr[1]=4.5;. This makes the program much easier to debug.

Another Dangling Pointer

A second way a dangling pointer may occur, and one that is more difficult to detect, is when two or more pointers point to the same memory location (i.e., they are aliases of the same storage) and the space is deallocated. For example, in the following code fragment

```
float *ptr = new float;
float *pptr = ptr;
// other statements here
delete ptr;
cout << *pptr; //undefined
```

`ptr` and `pptr` both point to the same storage location. When `ptr` is deleted, `pptr` becomes a dangling reference, and the same two problems discussed above may occur. To avoid this problem use aliases very carefully and only when necessary. (Solutions such as reference counters often work well but are beyond the scope of this text.)

Garbage and Memory Leaks

Garbage is a term used to refer to memory that has been allocated but never returned to the free store by a `delete` operation, and is no longer referenced by any pointer. Thus there is no way to access the memory, and it is "lost" to the program. This is also called a *memory leak*. It can occur, for example, by the following statements:

```
float *ptr = new float;
ptr = 0;
```

The memory allocated in the first line is garbage once `ptr` is set to 0. The statement `delete ptr;` should have been performed before `ptr = 0;`. Note the symmetry here with the dangling reference problem: If we `delete` memory but do not reset the pointer, we get a dangling reference. If we reset the pointer but do not `delete` the memory, we get a memory leak.

Another example of a memory leak is illustrated by the following:

```
float *ptr = new float;
//other statements here
*ptr = new float;
```

Here, memory is first allocated for a pointer to a `float`. Later in the program another memory allocation occurs. At this point, there is no longer a pointer to the first memory location allocated, and this memory is lost to the program. Again, `delete ptr;` should be executed before the second allocation.

A BIG Memory Leak

In the above examples, only the storage for single variables is lost, and this is not really a problem. But it can become a serious problem when memory is allocated and becomes garbage in a loop, as in the following example:

```
float *vec;
for (int i = 0; i < LARGE_VALUE; i++) {
  vec = new float[1000];
// some code using vec but not deleting vec
}
```

This program fragment will consume four thousand bytes of storage on every iteration, quickly exhausting memory.

16.2 MATRICES AND STRINGS

Matrix-Vector Multiplication

Using Pointer to Pointers

As an example of using the new operator with two-dimensional arrays, we give in Example 16.5 another version of a matrix-vector multiplication function. As opposed to the function in Example 15.4, A in the matvec() function in Example 16.5 is a pointer to pointers, as indicated by **A in the function heading. Likewise, in main(), A is defined to be a pointer to pointers rather than an array. The next statement uses the new operator to define storage for an array of pointers A[0], ..., A[n-1]; then the subsequent for loop assigns storage for each A[i], considered to be pointer to row i of the matrix. This is illustrated in Figure 16.2. Then A[i][j] is element j in row i. Since A[i] is also the name of the array holding row i, we may use array notation with it, as discussed in Section 15.2. Thus, by defining A as a pointer to pointers, we can use the usual array notation, A[i][j], in both the function matvec() and in main().

EXAMPLE 16.5
Matrix-Vector Multiplication Using Pointers to Pointers

```
#include <iostream.h>
void matvec(float **A, float *b, float *x, int n);
int main(){
  int n, i;
  cin >> n;
  float *b = new float[n];
  float *x = new float[n];
  for(i = 0; i < n; i++) x[i] = i;
  float **A ; //A is a pointer to pointers
  A = new float*[n]; //allocate storage for n pointers
  for(i = 0; i < n; i++)    //allocate storage
    A[i] = new float[n];    //for the matrix
  for(i = 0; i < n; i++)
    for(int j = 0; j < n; j++)
      A[i][j] = i + j;
  matvec(A,b,x,n);
  cout << "b[0] = " << b[0] << " b[1] = " << b[1] << endl;
  return 0;
}
void matvec(float **A, float *b, float *x, int n){
  for(int i = 0; i < n; i++){
    float temp = 0;
    for(int j = 0; j < n; j++)
        temp = temp + A[i][j] * x[j];
    b[i] = temp;
  }
}
```

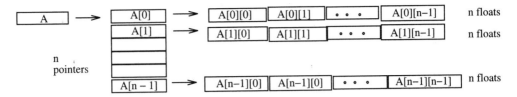

Figure 16.2 Pointer to Pointers to Array

The `for` loop following the memory allocation assigns numerical values to each element of the matrix, and could be replaced by any mechanism to assign values to A, for example, using `cin` statements. In the call to the function `matvec()`, A itself is the argument. This is consistent with the function heading, since A is a pointer to pointers.

A similar change may be made to the function `gauss()` of Example 13.1, so as to be able to handle linear systems of any size (see Exercise 16.4).

Strings

We next consider dynamic memory allocation in conjunction with strings. In Example 15.6 we gave a program in which we used an array of strings defined by

$$\text{char s[100][21]} \tag{16.2}$$

This defines an array of 100 strings, each of 20 characters plus the termination character. In Example 15.6, we used such an array of strings to hold the words of a sentence, each string holding one word. But this had a disadvantage: The number of characters in each string must be the same so that memory is wasted whenever a word is shorter than 20 characters. We can circumvent this problem by using an array of pointers to strings, rather than an array of strings. (The pointers, of course, also require storage.) A simple program that uses an array of pointers to strings is given in Example 16.6.

EXAMPLE 16.6
An Array of Pointers to Strings

```
#include <iostream.h>
int main(){
    char *sen[4] = {"I", "am", "a", "baker"};
    for(int i = 0; i < 4; i++)
        cout << sen[i] << endl;
    return 0;
}
```

Description of
Example 16.6

The first statement in the body of `main()` in Example 16.6 defines an array of pointers to strings, and initializes the strings as indicated. The strings themselves are stored contiguously in memory, and the pointers to those strings are stored in another array. This example does not use dynamic memory allocation; the lengths of the strings are determined by the initialization. Moreover, the number of strings is fixed at 4.

The next example, Example 16.7, is a modification of the program of Example 15.6 and overcomes both limitations of that program: We can work with an "arbitrary" number of strings of "arbitrary" length. Actually, the string lengths are not quite arbitrary. The first two statements in `main()` define a *buffer* of 99 characters plus the termination character, and this will be the maximum allowable string length.

EXAMPLE 16.7

Replacing Strings Using Dynamic Memory

```
//This program searches a "sentence" of n "words", each
//word up to 99 characters long and replaces a test string by
//another string assumed to be of the same length.
  #include <iostream.h>
  #include <string.h>
  int main(){
    int const MaxStringLength = 100;
    char buffer[MaxStringLength]; //set up a buffer
    cout << "What is the test string?\n";
    cin >> buffer; //input test string
    int len = strlen(buffer); //length of test string
    cout << "len =" << len << endl; //display string length
    char *test = new char[len + 1]; //allocate storage
    strcpy(test,buffer); //copy buffer
    char *replace = new char[len + 1]; //allocate storage
    cout << "what is the replacement string?\n";
    cin >> replace; //assumed same length as test string
    cout << "how many words?\n";
    int n;
    cin >> n;
    char **s; //declare s as pointer to pointers
    s = new char*[n]; //allocate storage for n pointers
    for(int i = 0; i < n; i++){ //loop for n words
    cout << "read a word\n";
    cin >> buffer;
    int len_s = strlen(buffer); //length of word
    s[i] = new char[len_s + 1]; //set storage for word
    strcpy(s[i], buffer); //copy buffer
    cout << "len_s=  " << len_s << "\n";
    cout << "input word is " << s[i] << endl;
    for(int j = 0; j <= len_s - len; j++) //loop through word
    fi(strncmp(test, &s[i][j], len) == 0) //string search
          strncpy(&s[i][j],replace,len); //replace string
    cout << "new string is " << s[i] << endl;
  }
    return 0;
  }
```

Buffer

But it is easily changed by the parameter MaxStringLength. We next read the test string into the buffer and determine the length of the string using the `strlen()` function of (15.18). Knowing the length of the test string, we can then dynamically allocate storage for it. Note that we cannot allocate storage until we know the length of the string, and this is the purpose of `buffer`: It is a place to store temporarily this and other strings. Note also that `len+1` bytes of storage are allocated to have space for the termination character. We then transfer the contents of the buffer to the storage locations allocated for `test`. An important point here is that we can *not* do this transfer by a simple assignment statement such as `test = buffer`. Since `test` is a pointer, this would assign to `test` the address of `buffer`, and henceforth `test` would be an alias for `buffer`. Instead we use the function `strcpy()` of (15.21) to effect the transfer. Since we are assuming that the replacement string, `replace`, is the same length as the test string, we can allocate storage for it and read the replacement string directly into this space without using the buffer.

Transferring from Buffer

Next we read `n`, the number of words. We then define `s` as a pointer to pointers and allocate storage for `n` pointers. This is the same construction used in the matrix-vector multiplication program of Example 16.5. We now begin a loop to examine `n` words. Since we do not know the length of these words, we again use the buffer and proceed exactly as was done with `test`. After a word `s[i]` has been read, we begin a loop to try to match the `test` string with a substring of `s[i]`. This is exactly the same loop as used in Example 15.7. As in Example 16.5, we can use the array notation `s[i][j]` in conjunction with `s` being a pointer to pointers. An important point in this loop is that we must use the function `strncpy()` rather than `strcpy()`, since we wish exactly `len` characters to be copied.

16.3 LINKED LISTS

We next give one further, and important, example of the use of pointers and dynamic memory allocation. Many vectors and, especially, matrices that arise in scientific computing have mostly zero elements. For example, suppose that x is a 500-long vector but only the elements x_4, x_{35}, x_{121}, and x_{365} are nonzero. Such a vector is called *sparse*. To store this vector in a 500-long array would be very wasteful of storage since most elements of the array will be zero. All that is really needed is to know the values of the nonzero elements and their location in the vector. We could keep this information in two short arrays, as indicated in Figure 16.3.

Figure 16.3 A Sparse Vector Stored as Two Arrays

Index Array
4
35
121
365

Value Array	
4.215	(value of x_4)
3.142	(value of x_{35})
1.426	(value of x_{121})
2.819	(value of x_{365})

Representing a sparse vector by two arrays, as illustrated in Figure 16.3, does have some problems, however. As the vector is operated on, perhaps by addition to another sparse vector, the number and locations of its nonzero entries may change. For many purposes, it is useful to keep the indices in ascending order. Thus, if a new entry, say with index 89, is made, all the elements of the two arrays beyond that index must be moved accordingly. (The array sizes may also change, but we can handle that by dynamic memory allocation.) An alternative very useful representation is called a (*singly*) *linked list* and is illustrated in Figure 16.4.

Each block in Figure 16.4 has three things: The index, the value of the vector entry corresponding to that index, and a pointer that gives the address of the next block in the list. Such blocks are called *nodes* in the list, and can be represented quite naturally by structures (Chapter 11). Thus we could define a structure as shown in Example 16.8.

EXAMPLE 16.8
A Node Represented by a Structure

```
struct node{
  node *next;
  int index;
  float value;
};
  node Node4 = {0, 365, 2.819};
  node Node3 = {&Node4, 121, 1.426};
  node Node2 = {&Node3, 35, 3.142};
  node Node1 = {&Node2, 4, 4.815};
  node *head = &Node1;
```

In the `struct` of Example 16.8, `next` is defined as a pointer to a `node`, and then `index` and `value` are the other two members of the structure. In order to establish a linked list, we could just define nodes as `struct`s with the desired indices, values, and pointers. For example, the linked list of Figure 16.4 could be defined by the statements shown in Example 16.8. Here, we have represented the `End` pointer by the null pointer 0, and we define the last node first with this 0 value for the pointer. Then `Node3` is defined with its pointer set to `&Node4`, the address of `Node4`, and so on. Finally, the pointer `head` is set to the address of `Node1`.

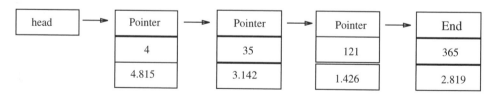

Figure 16.4 A Linked List

There are many operations that can be performed on linked lists, and we will write functions for a few of the most basic ones. We first give in Example 16.9 a function for checking to see if a given index is already in the list. If it is, the function returns a pointer to the node containing that index; thus the return value of the function is a pointer to a node, as indicated by `node*` as the function type. We assume that the pointer in the last node in the list (the end marker in Figure 16.4) is 0, as used in Example 16.8. The second argument of the function `member()` in Example 16.9 is `ind`, and we wish to see if a node with an index of `ind` is in the list. The other argument of the function is the pointer `head`; thus this function can be used on any other linked list of the same form, which will be useful if we are representing several sparse vectors by linked lists. The first statement of the function body defines a pointer, `current`, to nodes and initializes it to the value of `head`; that is, `current` is initialized to the pointer to the first node in the list. We then start a `while` loop, which terminates when `current == 0`, which signifies the end of the list. The arrow operator used in the `while` loop combines the dot operator to reference members of a `struct` and the deferencing operator `*`; that is,

EXAMPLE 16.9
A Function to Scan a Linked List

```
node* member(node* head, int ind){
//Scan the list checking if ind is an index in the list
   node *current = head;
   while(current != 0){
     if(current->index == ind) return current;
     current = current->next;
   }
return 0; //We did not find ind in the list, so return 0
}
```

current -> index   is same as   (*current).index

The -> Operator   Here, since `current` is a pointer to a node, `*current` is the node itself and so `(*current).index` is the index member of that node. If the index value of the current node (`current->index`) is `ind`, the function returns the pointer to that node. If the `while` loop terminates without finding `ind`, the function returns the null pointer 0. Thus if you are only interested in whether or not an index is in the list, a test of the function return value (0 or nonzero) gives that information. But the location of the node with that index is also available.

Printing a List   The next function we give is one to display a list, and is very similar to the one for scanning a list. The `PrintList` function shown in Example 16.10 displays the indices and values of the nodes of a list (it wouldn't make sense to display the pointers since these are binary addresses). For example, if this function is called for the linked list of Example 16.8, the output will be

EXAMPLE 16.10
A Function to Display a List

```
//This function displays a list whose beginning address is head
void PrintList(node* head){
  node *current = head;
  while(current != 0){
    cout << current->index << " "
         << current->value << endl;
    current = current->next;
  }
}
```

```
    4   4.815
   35   3.142
  121   1.426
  365   2.819
```

Adding a Node

The last (and most difficult) function we will give is one to add a node to a list. For example, suppose we wish to add a node with index 41 and value 4.921 to the linked list of Figure 16.4. Then we need to change the pointer in the previous node to point to the new node and set the pointer in the new node to point to the next node in the list. This is illustrated in Figure 16.5.

Example 16.11 gives a function to add a new node to a list in such a way that the indices of the new list are still in increasing order, as shown in Figure 16.5. We first define a function, CreateNode(), that allocates memory for a new node and sets the values of its member variables. As in Example 16.9, the return type of this function is a pointer to a node, which will be a pointer to the new node. This function is used in the following function, AddNode(), which assumes for simplicity that the node to be added has an index that is not equal to any index already in the list. (Exercise 16.9 asks you to modify AddNode() so as to replace the value of a node, rather than add a new node, if the index equals one already in the list.) The arguments of AddNode() are a pointer to the first node of the list, and the index and value of the new node.

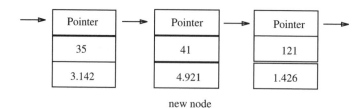

new node

Figure 16.5 Adding a New Node

EXAMPLE 16.11
Adding a New Node to a List

```
                    //This function establishes a new node
                  node* CreateNode(node* ptr, int ind, float val){
                    node *NewNode = new node; //allocate memory
                    NewNode->next = ptr;        //set pointer
                    NewNode->index = ind;       //set index
                    NewNode->value = val;       //set value
                    return NewNode; //pointer to node is returned
                  }
                  //This function adds a node to the list with beginning address
                  //head.  It is assumed that the index of the new node differs
                  //from all indices already in the list
                  void AddNode(node* &head, int ind, float val){
                     node *current = head;
                     node *last = head;
                     if(head == 0){ //if list is empty, set head to new node
                       head = CreateNode(0, ind, val);
                       return;
                     }
                     if(ind < current->index){ //new first node
                       node *NewNode= CreateNode(head, ind, val);
                       head = NewNode; //head now points to new node
                       return;
                     } //new node is not first. find its place by algorithm
                     //similar to that in function member()
                     while(current != 0)
                       if(ind > current->index){
                         last = current; //save current pointer
                         current = current->next; //go to next node
                       }
                       else{ //add new node
                         node *NewNode= CreateNode(current, ind, val);
                         last->next = NewNode; //set pointer to new node
                         return;
                       }
                  //If we get to here, new node becomes last one in list
                     node *NewNode = CreateNode(0, ind, val);
                     last->next = NewNode;
                  }
```

We first check to see if head is the null pointer, which implies that the list is empty. If so, we establish a new node with the null pointer as its pointer value; this is then the only node in the new list. Next, there are three cases to consider: the new node is the first in the new list, in the middle of the list, or the last in the new list. The first if statement handles the first case; here, head is changed to point to the new node and this new value of head is returned implicitly by the function. Note

head *Called by Reference*

that in AddNode(), as opposed to the previous functions, head is called by reference so that the new value of head may be returned. If this were not done, the value of head would not be changed outside the function and the new first node could not be accessed.

If the new node is not the first one in the list, a `while` loop moves through the list, as was done in Example 16.9. The test in the `if` statement now asks whether `ind` is greater than the index of the current node. If it is, we move to the next node; otherwise, we insert the new node at this point. The function `CreateNode()` establishes this new node with a pointer to the next node in the list. Then we set the current pointer to point to the new node. In order to do this, we have established an auxiliary pointer variable `last` to hold the current pointer before it is changed. Then when a new node is established, the statement

Adding Nodes in the Middle

```
last->next = NewNode:
```

sets the pointer of the node before `NewNode` to point to `NewNode`. (We note that with a *doubly* linked list, each node has a pointer to both the next and the previous nodes. In this case, changing the pointer of the previous node is a little easier.) If we exhaust the `while` loop, this means that the index of the new node is greater than any index in the list; so the new node should become the last node in the list, with a pointer value of 0.

Example 16.12 gives a function `main()` that illustrates the use of the `AddNode()` and `PrintList()` functions. This program first establishes the list shown in Figure 16.4 by successively adding nodes to the first node, and then adds another node in the middle of the list.

EXAMPLE 16.12

Using the `AddNode()` and `PrintList()` Functions

```
//This program calls the functions AddNode() of
//Example 16.11 and PrintList() of Example 16.10
int main(){
  node *head = 0; //initialize head
  AddNode(head,4,4.815); //Set first Node in list
  AddNode(head,35,3.142); //Add three more nodes
  AddNode(head,365,2.819);
  AddNode(head,121,1.426);
  PrintList(head); //Display list
  AddNode(head,1,9.786); //Add another node
  PrintList(head); //Display new list
  return 0;
}
```

MAIN POINTS OF CHAPTER 16

- Memory may be allocated dynamically by the `new` operator, and should be deleted by the `delete` operator when no longer needed.
- Memory dynamically allocated within a function persists outside the function, as opposed to memory allocated statically within a function.

- Two common problems with dynamic memory allocation are dangling references (dangling pointers) and garbage (memory leaks). Both may be overcome by proper deletion of dynamic memory.
- Matrix computation may be done effectively by defining the matrix as a pointer to pointers, rather than an array, and then dynamically allocating storage for the pointers as well as the elements.
- Dynamic memory allocation may be used in conjunction with strings and allows the size of the strings to determine the memory needed. Again, the use of a pointer to pointers is an effective way to handle the memory.
- Linked lists are an important data structure that may be implemented by the use of pointers and dynamic memory allocation.

EXERCISES

16.1 Write a `main()` program that reads an integer n, dynamically allocates memory for two arrays a and b of length n, sets values for the array elements, and then displays the sum of the arrays.

16.2 Write a `main()` program that calls the function of Example 16.3 and then tries to use the array b. Verify that there are problems. Then call, instead, the function of Example 16.4 and verify that there are no problems.

16.3 Run the program of Example 16.5 for various values of n, modifying the array x suitably. Then modify the program so that the function computes the sum of all elements of a two-dimensional array, rather than a matrix-vector product. Modify `main()` accordingly.

16.4 Modify the function `gauss()` of Example 13.1 so that the function parameters are the same as those of Example 16.5. Then write a `main()` program to test this function for systems of different sizes. Use both dynamic and static memory allocation for the array A in `main()`.

16.5 Run the program of Example 16.7 for various values of the `test` and `replace` strings and various words. Try to modify the program so that the `test` and `replace` strings do not have to be the same length.

16.6 For the linked list of Figure 16.4, as implemented in Example 16.12, use the function `member()` of Example 16.9 to determine if certain nodes are in the list. Then modify the function `member()` so that it checks the `value` of a node, rather than the index. Comment on what difficulties you might expect with the modified program because of rounding error.

16.7 Change the function `main()` of Example 16.12 so as to establish a different linked list. Then add nodes to the list at the beginning and end. Verify that the functions `AddNode()` and `PrintList()` are working correctly for your linked list.

16.8 Define a linked list of integers kept in descending order. Modify the functions of Examples 16.9, 16.10, and 16.11 to operate on such a linked list.

16.9 Use the function `member()` of Example 16.9 to modify the function `AddNode` of Example 16.11 so as to replace the `value` of a node, rather than add a new node, if the `index` is already in the list.

PART III

OBJECT-ORIENTED PROGRAMMING

In the first two parts of this book, we have covered most of the C++ language with the very important exception of classes. In fact, with a few more or less minor differences, what we have covered so far is essentially the language C, the predecessor of C++. Now, in this final part, we cover classes and their use in object-oriented programming. It is classes that primarily sets C++ apart from C.

Chapter 17 gives a basic introduction to classes and objects, including the concepts of information hiding and abstract data types. We cover many of the technical details of defining and using classes including constructors, friend functions, and operator overloading. We end this chapter with a discussion of stream classes, which puts into proper perspective our previous use of `cin` and `cout`, and other input/output statements.

Chapter 18 covers the use of dynamic memory in classes, which gives rise to a number of potential problems requiring copy constructors and other operators. The main focus of this chapter is the definition of classes for arrays, both for vectors and matrices, which are very important in scientific computing.

Finally, in Chapter 19 we treat the important mechanism of inheritance, which allows us to derive new classes from existing classes. We end this chapter with an example of a linear equation solver that automatically chooses the right form of Gaussian elimination to use, depending on properties of the coefficient matrix.

chapter
17

CLASSES AND OBJECTS

In this chapter we will introduce classes and objects and how to use them in what is called *object-oriented programming*. In Chapter 11, we saw how to define new data types by means of structures. A *class* is another user-defined data type, also called an *abstract data type*, that is defined in terms of fundamental data types, used-defined data types, functions, and other classes. An *object* is a particular instance or *instantiation* of a given class. In addition to allowing the user to define new data types, classes and objects have many of the same advantages as functions:

- Classes help to manage the complexity of large programs by allowing modularity, that is, dividing the program into smaller pieces that can be separately compiled and tested.
- Classes facilitate software reuse.
- Classes contribute to fault containment; that is, errors may be localized within the class.
- Classes allow *information hiding*, which means that functions and data structures within the class are not known or available to users of the class and hence may be changed without affecting users adversely.

17.1 A SIMPLE CLASS

As a first example of a class, we define in Example 17.1 a class of complex numbers. The code in Example 17.1 is called the *class specifier* or *class definition*. The first line gives the class name: `complex`. We next declare two *member variables* of type `float` that will be the real and imaginary parts of a complex number. If we stopped at this point, we would have something like a structure of complex numbers, as defined in Chapter 11 (see Exercise 11.9). However, classes may also contain *member functions*. (Actually, the C++ `struct` construct also permits member functions, but `structs` are rarely used in C++ in this way because the class construct is more general.)

The first function in Example 17.1, `set()`, sets values of `real` and `imag` and is called a *manipulator* or *mutator* member function since it changes data. The second and third member functions, `real_part()` and `imag_part()`, are *accessor* or *inspector* functions since they return the variables `real` and `imag`. We will see how these three member functions are utilized in a moment. Note that there must be a semicolon following the brace that ends the class definition.

EXAMPLE 17.1
A Class for Complex Numbers

```
class complex{
  private:
    float real; //real part of complex number
    float imag; //imaginary part of complex number
  public:
    void set(float c,float d){ //manipulator function
      real = c; //set real part
      imag = d; //set imaginary part
    }
    float real_part(){return real;} //accessor function
    float imag_part(){return imag;} //accessor function
};
```

Information Hiding

The variables `real` and `imag` in Example 17.1 are declared to be `private`. This achieves *information hiding*, which means that these variables may be used only by member functions of the class, and are not directly available from outside the class. Private variables defined within a class are said to be *encapsulated* and have *class scope*; that is, they are directly accessible only within the class. Thus they are somewhat like local variables within a function.

public *versus* private

The functions in Example 17.1 are declared to be `public`, which means that they are accessible from outside the class and provide the interface to the class. In general, variables and functions may be declared to be either `private` or `public`, but it is good practice for variables to be private. By default, all class members are private unless declared otherwise so that it is not necessary to use the word `private` explicitly. However, for clarity we will continue to do so. The key words `private` and `public` are called *class access specifiers*. (We note that a `struct` with member functions is just a class in which all members are public.)

We next give in Example 17.2 an example of how the class of Example 17.1 may be used. In `main()`, we first declare or *instantiate* `alpha`, an *object* of the class `complex`. We then call the function `set()` of the class to give values to the real and imaginary parts of the object `alpha`; note that we call this function by appending the function name, `set()`, to the object name `alpha`:

EXAMPLE 17.2
Using the Class Complex

```
#include <iostream.h>
int main(){
    complex alpha;              //an object of type complex
    alpha.set(1.0,1.0);         //set value of complex number
    cout << alpha.real_part() << alpha.imag_part() << endl;
    return 0;
}
```

`alpha.set(1.0,1.0)`

Calling Member
Functions

The dot operator in this construction is called the *class member access operator*. (Recall that the dot operator is also used to access members of a structure.) There is no reference to the private variables `real` and `imag` of the class, but 1.0 and 1.0 are now the real and imaginary parts of the complex number associated with `alpha`. Thus, when the functions `real_part()` and `imag_part()` are called, it is these values of `real` and `imag` that are returned. Note that we cannot access the private variables `real` and `imag` directly. For example, use of `real` or `alpha.real` in `main()` would be illegal and cause a syntax error.

A Better Complex Class

The class in Example 17.1 is not yet very useful, and we will rewrite it so that it contains functions that do addition and multiplication of complex numbers. (Exercise 17.2 asks you to add a member function to do division of complex numbers.) This new class will also illustrate several other aspects of classes and objects.

If

$$\alpha = a + ib, \quad \beta = c + id$$

are two complex numbers, where $i = \sqrt{-1}$ as usual, the rules for addition and multiplication of α and β are

Complex Arithmetic

$$\alpha + \beta = (a + c) + i(b + d) \tag{17.1a}$$

$$\alpha\beta = (a + ib)(c + id) = (ac - bd) + i(ad + bc) \tag{17.1b}$$

Thus, if $\gamma = \alpha + \beta$, then from (17.1a) the real and imaginary parts of γ are

$$e = a + c, \quad f = b + d \tag{17.2}$$

Similarly, if $\gamma = \alpha\beta$, from (17.1b) the real and imaginary parts of γ are

$$e = ac - bd, \quad f = ad + bc \tag{17.3}$$

For information hiding, it is usually best to give only function prototypes within the class definition and define the functions themselves later; this is the approach we use now. We will put these prototypes in a separate header file (see Section 14.4) called Complex.h that will also contain the rest of the class definition. This header file is shown in Example 17.3 and gives prototypes for the public functions complex(), read(), print(), add(), and mult( ). (The two "constructor" functions will be discussed in the next subsection.)

EXAMPLE 17.3
Header File complex.h

```
#include <iostream.h>
class complex{
   private:
     float real; //real part of a complex number
     float imag; //imaginary part of a complex number
   public:
     complex() {}  //default constructor
     complex(float, float); //constructor prototype
     void read();  //prototype to read real and imag
     void print(); //prototype to print real and imag
     complex add(const complex&);   //addition prototype
     complex mult(const complex&); //multiplication prototype
};
```

EXAMPLE 17.4
Implementation File for the Complex Class

```
complex :: complex(float re, float im){ //constructor
   real = re;
   imag = im;
}
void complex :: read(){
   cout << "enter real and imag parts\n";
   cin >> real >> imag;
}
void complex :: print(){
   cout << "real = " << real << " imag = " << imag << endl;
}
complex complex :: add(const complex &beta){
   complex delta;
   delta.real = real + beta.real;
   delta.imag = imag + beta.imag;
   return delta;
}
complex complex :: mult(const complex &beta){
   complex delta;
   delta.real = real * beta.real - imag * beta.imag;
   delta.imag = real * beta.imag + imag * beta.real;
   return delta;
}
```

Defining Functions Outside the Class

The member functions are defined in a separate *implementation* or *source* file, as shown in Example 17.4. Note that in the definitions of these member functions the first line is

```
void complex :: read()
```

where `void` is the return type and `complex` is the class name, and similarly for the other functions; this is the syntax needed to define member functions of the class `complex` outside of the class definition. The operator : : is called the *scope resolution operator*.

Description of Example 17.5

Example 17.5 illustrates use of the class `complex` of Examples 17.3 and 17.4. The function `main()` in Example 17.5 declares three objects of the class `complex`: `alpha`, `beta`, and `gamma`. The real and imaginary parts of `alpha` are set to 1.0 by means of the constructor function, to be discussed in the next subsection. Then the statement `beta.read()` calls the member function `read()`, and the real and imaginary parts of a complex number are entered at the keyboard. Note that `read()` has no arguments, but after the function call is complete, the real and imaginary parts of the `private` complex variable `beta` have been set.

EXAMPLE 17.5
Using the Class Complex

```
int main(){
  complex beta,gamma; //declare two objects
  complex alpha(1.0,1.0); //declare and initialize object
  beta.read(); //read complex number beta from keyboard
  gamma = alpha.add(beta); //gamma = alpha + beta
  gamma.print();
  gamma = alpha.mult(beta); //gamma = alpha * beta
  gamma.print();
  return 0;
}
```

Next, `gamma` is assigned the sum of `alpha` and `beta`. In the definition of the function `add()`, the single argument `beta` is called by reference, and `const` ensures that `beta` cannot be changed by the function `add()` (see Chapter 14). Recall from Chapter 14 that an advantage of calling by reference is that a copy of the argument does not need to be made within the function. In general, it is more efficient for object arguments of a function to be called by reference rather than by value, and we will uniformly do this. Since the member function `add()` is invoked on `alpha`, it is `alpha` that is implicitly the other argument of this function; that is, in the statement

```
delta.real = real + beta.real;
```

in the definition of `add()`, `real` will be the real part of `alpha` (and similarly for the imaginary part).

After the addition of `alpha` and `beta`, the statement `gamma.print()` then calls the member function `print()` to display the real and imaginary parts of `gamma`. The process now repeats to obtain the product of `alpha` and `beta`. Note that both `add()`

Classes Are
First-Class Types

and `mult()` return an object, not just a single variable, since the first `complex` in the first line of the function definitions of `add()` and `mult()` in Example 17.4 says that the return type of the function is `complex`. More generally, a class is a first-class type so that objects may be used as function parameters as well as return values.

We note that instead of using header and implementation files the program of Examples 17.3-17.5 could be written as one file with the statements of Example 17.3 followed by those of Example 17.4 and then those of Example 17.5.

Constructors

Can't Initialize
Member Variables
in Class

When we declare a variable of fundamental type, we may also initialize it, as in

$$int\ i = 3;$$

However, we cannot initialize member variables within a class in this way. For example, a statement like `float real = 1.0` in the definition of the `complex` class of Example 17.3 would not be legal. Moreover, we will usually wish to do initializations not within the class but when we declare an object. This may be accomplished by a special type of member function called a *constructor*, as is illustrated in Examples 17.3 and 17.4.

The name of a constructor function is always the class name, and a constructor function has no return type, not even `void`. The first function defined in Example 17.4 is a constructor. This function sets the private variables `real` and `imag` to the values `re` and `im`, and is invoked in `main()` when the object `alpha` is declared. Thus the private variables `real` and `imag` of `alpha` are initialized to 1.0 and 1.0. This is a much better mechanism for initialization than the use of a manipulator function like `set()` in Example 17.2. Two other objects, `beta` and `gamma`, are also declared in `main()` but with no initial values given. The private variables of these objects are also initialized, but by the *default constructor*, which is the first public member function in Example 17.3. This default constructor is parameterless (it has no arguments), and has no function body. (Since it is so simple, we have put the entire function, including its implementation, in the header file; thus it does not appear in Example 17.4.)

Default
Constructor

Constructors did not appear in the class of Example 17.1 because no objects were explicitly initialized by the program in Example 17.2. However, a default constructor was automatically supplied by the compiler when an object was declared. When a constructor is defined, as in Example 17.4, it is sometimes necessary to furnish the default constructor explicitly. In particular, if some objects are declared without initialization, as are `beta` and `gamma` in Example 17.5, but other objects are declared with initialization by a constructor, as is `alpha` in Example 17.5, then an explicit default constructor must be given, or else syntax errors will occur (see Exercise 17.3).

Default Parameters

Constructors may also use default parameters (Section 14.3). For example, the constructor in Example 17.4 could have been written as

```
complex(float re, float im = 0) {real = re; imag = im;}
```

with a default value of 0 given to `im`. Then, the declaration

```
complex alpha(1.0);
```

in `main()` will set the real part of `alpha` to 1.0, and the imaginary part takes its default value of 0. But we can always override the default value if we choose. Thus the statement

```
complex alpha(1.0, 2.0);
```

would set the imaginary part to 2.0.

Default constructors always have no parameters, but they do not necessarily need to have an empty function body, as in Example 17.3. For example, we could change the default constructor to

Default
Constructors with
Body

```
complex() {real = 1; imag = 0;}
```

Then the statement

```
complex beta, gamma;
```

in `main()` will use this default constructor to initialize `beta` and `gamma` to have real part equal to 1 and imaginary part equal to 0.

Since constructors are an essential part of defining classes and are initially rather confusing, we give here a summary of when they must be used.

Summary of
Constructors

- A constructor is used to initialize objects when they are declared. This constructor may contain parameters that set initial values of one or more of the private variables of the object to the values given by the parameters. Default parameters may also be used.

- A default constructor is used automatically by the compiler when an object is declared and no values of member variables are specified. A default constructor does not need to appear explicitly in the class definition unless an explicit constructor is also used. A default constructor has no parameters and usually no body, although a body may be supplied to give specific initial values to one or more of the member variables.

Arrays of Objects

In Chapter 11, we noted that we can define arrays of `structs`, and we may also define arrays of objects. Suppose, for example, that we wish to define an array of complex numbers. If the class `complex` has been defined as in Examples 17.3 and 17.4, in `main()` we may define arrays of complex numbers by

```
complex alpha[10], beta[10], gamma[10]; //arrays of objects
```

Then we might have somewhere in `main()` the loop

```
for(i = 0; i < 10; i++)
    gamma[i] = alpha[i].add(beta[i]);
```

which would add the arrays `alpha` and `beta`, putting the results into the array `gamma`.

Exercise 17.4 asks you to modify the program of Examples 17.3–17.5 to use arrays of complex objects.

Initialization of Arrays

Arrays of objects cannot be initialized by an explicit constructor and are initialized by the default constructor. If an explicit constructor is included in the class definition for other purposes, then an explicit default constructor must be included also. This is another instance of when an explicit default constructor is necessary.

const **Objects and Functions**

As with other data types, class objects may also be declared to be `const`. For instance, in Example 17.5, we could declare the object `alpha` by

```
const complex alpha(1.0, 1.0);
```

Member functions may also be given the qualifier `const`, as illustrated by the following prototype for the function `add()` of Example 17.3:

```
complex add(const complex &) const;
```

`const` would appear in the same place in the function definition in Example 17.4. A member function without the `const` qualification cannot be used on a `const` object. For example, if `alpha` is declared `const`, as above, then `add()` must have the `const` qualifier in order that the statement

```
gamma = alpha.add(beta);
```

be valid. (On some compilers, there may be a warning rather than a syntax error.) On the other hand, functions with the `const` qualifier may also be used on non-`const` objects.

Conditional Compilation[†]

In Section 3.3, in the context of debugging a program, we discussed conditional compilation that is controlled by the directives `#if` and `#endif`. A similar mechanism may be desirable when defining classes, in order to guarantee that no class is erroneously defined more than once. To this end, the header file `complex.h` of Example 17.3 could be written as:

```
#ifndef COMPLEX_H
#define COMPLEX_H
#include <iostream.h>
class complex{
  //class definition here as in Example 17.3
};
#endif
```

[†]This subsection may be omitted on first reading.

Following the directive `#ifndef` is the name of the header file but written in capital letters with the . replaced by _. This name is repeated in the following `#define` directive. Then the code following the `#define` directive up to the `#endif` directive will be compiled only if the same class has not been previously defined. In the small programs of this book, there will be no occasion to use this conditional compilation mechanism, but for large programs, especially those written by several people, this mechanism is useful in order to prevent inadvertent multiple definitions.

Static Members[†]

If the statement

$$\text{int n;}$$

Static Variables
defines a (private or public) variable in a class, then every object of the class has its own value of n. However, if the declaration of n is

$$\text{static int n;} \tag{17.4}$$

then n will have the same value in all objects of the class. In particular, there will be only one copy of n, no matter how many objects are declared. Such a variable is said to be *static*. (Recall that static variables in functions were discussed in Section 14.3.) Static variables may also be initialized outside the class if they have been declared `public`; for example,

$$\text{ClassName :: n = 0;}$$

might appear in `main()`. Note that the class name followed by the scope resolution operator must be used here.

Member functions may also be static; for example,

$$\text{static int count(int n) \{return n++\};} \tag{17.5}$$

Static Functions
would define a function that increments the static variable n, and after this function is executed the effect is that every object has access to the incremented value of n. Static member functions can only access static data members. Static member functions may be called in the usual way, but they may also be called independently of a particular object. For example, if (17.4) and (17.5) are definitions in a class named `large`, then the statement

$$\text{large :: count(n)}$$

in `main()` calls the function `count()`.

17.2 CLASSES AND FUNCTIONS

In the previous section we introduced classes and member functions of classes. In this section, we discuss some nonmember functions that can be used with

---

[†]This subsection may be omitted on first reading.

classes and then see how we can define member functions that "overload" operators such as + and * so that they can be used on objects like those defining complex numbers.

Auxiliary Functions versus Member Functions

The class in Examples 17.3 and 17.4 contained two member functions, add() and mult(), that do arithmetic operations on complex numbers. Alternatively, we could put these functions outside the class as *auxiliary functions*, an approach preferred by some writers. The class complex defined in Example 17.6 now contains all of the member functions of both of the classes of Examples 17.1 and 17.3, except the functions add() and mult(), which are defined outside the class. We have also added more manipulator functions for setting the real and imaginary parts separately.

EXAMPLE 17.6
Another Complex Class

```
class complex{
  private:
    float real; //real part of a complex number
    float imag; //imaginary part of a complex number
  public:
    complex() {} //default constructor
    complex(float re, float im){real = re; imag = im;}
    void read(); //prototype. function defined in Example 17.4
    void print();//prototype. function defined in Example 17.4
    float real_part() {return real;}
    float imag_part() {return imag;}
    void set_real(float c) {real = c;}
    void set_imag(float d) {imag = d;}
    void set(float c, float d) {real = c; imag = d;}
  };
```

Example 17.7 shows how the class of Example 17.6 may be used and also gives a definition of the function add(). It is left to Exercise 17.5 to give a corresponding definition of the function mult(). The definitions of read() and print() are not shown since they are the same as in Example 17.4.

Description of
Example 17.7

The program in Example 17.7 illustrates how the different member functions of the class of Example 17.6 can be used. As in Example 17.5, the constructor initializes alpha when it is declared. The manipulator function set() is used to give values to delta in main() and the functions set_real() and set_imag(), as well as real_part() and imag_part(), are used in the function add(). Note that accessor and manipulator functions like these are needed since the function add() is now not a member function of the class and has no direct access to the private variables `real` and `imag`.

EXAMPLE 17.7
An Auxiliary Function for the Class of Example 17.6

```
#include <iostream.h>
complex add(complex &alpha, complex &beta){
  complex delta;
  delta.set_real(alpha.real_part() + beta.real_part());
  delta.set_imag(alpha.imag_part() + beta.imag_part());
  return delta;
}
int main(){
  complex beta, gamma, delta; //declare three objects
  complex alpha(1.0, 1.0); //declare and initialize object
  beta.read(); //read complex number beta from keyboard
  gamma = add(alpha, beta); //add alpha and beta
  gamma.print();
  delta.set(2.0, -4.0); //set an object
  gamma = add(gamma, delta); //add gamma and delta
  gamma.print();
  return 0;
}
```

In most cases, we will prefer to use member functions rather than auxiliary functions in the rest of this book.

Friend Functions

In general, private members of a class may only be accessed by member functions of the class and not by functions outside the class, as we just saw in our discussion of auxiliary functions. However, a function may be declared to be a *friend* of a class and then it may access all private members of the class. We will illustrate this by a program to compute the absolute value of a complex number $a + ib$ by the formula $|a + ib| = \sqrt{a^2 + b^2}$. This program, using a friend function, is shown in Example 17.8.

Description of Example 17.8

As Example 17.8 shows, because `abs_value()` has been declared to be a friend function, it may access the private variables `real` and `imag`, just as if it were a member function, rather than having to use accessor functions as was done in the function `add()` of Example 17.7. Moreover, the absolute value may be computed in `main()` using the more mathematical function notation `abs_value(alpha)`, as opposed to the member function form `alpha.abs_value()` required if `abs_value()` had been a member function.

EXAMPLE 17.8
Use of a Friend Function

```
#include <iostream.h>
#include <math.h>
class complex{ //define a class
  private:
    float real, imag; //real and imaginary part
  public:
    complex() {}
    complex(float re, float im) {real = re; imag= im;}
    friend float abs_value(const complex&);//friend prototype
};
float abs_value(const complex &z){ //define friend function
  return sqrt(z.real * z.real + z.imag * z.imag);
}
int main(){
  complex alpha(1.0, 1.0);   //an object of type complex
  float abs_alpha;    //absolute value
  abs_alpha = abs_value(alpha); //compute absolute value
  cout << abs_alpha  << endl;
  return 0;
}
```

A similar change may be made to the function add() of Example 17.7. In this case, the prototype

```
            friend complex add(const complex&, const complex&);
```

Making add() *a*
Friend Function

would be put in the class and the implementation of the function would be

```
complex add(const complex &alpha, const complex &beta){
  complex delta;
  delta.real = alpha.real + beta.imag;
  delta.imag.= alpha.imag + beta.imag;
  return delta;
}
```

Exercise 17.6 asks you to install this add() function in Example 17.8. Note that, as illustrated by add() as well as abs_value() in Example 17.8, the key word friend goes before the function name in the prototype statement in the class definition but is not used in the definition of the function.

A function may not be a member function of two different classes but a friend function can circumvent this restriction since friend functions may access private data in two or more classes. This is illustrated schematically in Example 17.9. The friend function func() may access private data from class1 as its first argument and from class2 as its second argument, something that could not be done with just a member function.

EXAMPLE 17.9
A Friend Function for Two Classes

```
class class1{ //define a class
    private:
      //define variables here
    public:
      friend func(class1, class2); //prototype
};
class class2{ //define another class
    private:
      //define variables here
    public:
      friend func(class1, class2); //prototype
};
```

Friend Classes

If we have several functions in a class and we would like all of them to be friend functions of another class, rather than defining each individual function to be a friend function, we may define the whole class as a friend. For example, as in Example 17.9, suppose we have two classes class1 and class2. If in the public section of class1 we have the statement

```
                   friend class1 class2;
```

then every member function in class2 will be a friend function and may access any of the private members in class1.

Since friend functions tend to negate the benefits of information hiding, they should be used sparingly and only when necessary.

Overloading Operators

In Example 17.5 the statements

```
      gamma = alpha.add(beta); gamma = alpha.mult(beta);
```

in main() formed the sum and product of the complex objects alpha and beta. Although this is fairly readable, it would be much better to be able to write

```
      gamma = alpha + beta; gamma = alpha * beta;
```

However, the operators + and * are defined only for fundamental data types and not for objects. To use them for user-defined data types, we must define their meaning. This is called an *overloading* of an operator: It is assigned a meaning in addition to its usual ones for fundamental types. This is accomplished by defining another member function of the class to carry out the overloaded operation, and is illustrated in Example 17.10 for the class complex of Example 17.3. For brevity, we have eliminated the multiplication function from Example 17.3 (see Exercise 17.8), and we also do not show the code for the member functions read() and print(), which are the same as in Example 17.4.

EXAMPLE 17.10

A Complex Class with an Overloaded Operator

```
#include <iostream.h>
class complex{
   private:
      float real; //real part of a complex number
      float imag; //imaginary part of a complex number
   public:
      complex() {} //constructors
      complex(float re,float im)real = re; imag = im;
      void read(); //function prototype
      void print();//function prototype
      complex operator+ (const complex &); //operator prototype
};
//Define the overloaded operator +
inline complex complex :: operator+ (const complex &beta){
   complex delta;
   delta.real = real + beta.real;
   delta.imag = imag + beta.imag;
   return delta;
}
int main (){
   complex beta,gamma; //declare two objects
   complex alpha(1.0,1.0);//declare and initialize object
   beta.read(); //read defined in Example 17.4
   gamma = alpha + beta; //use overloaded operator
   gamma.print(); //print defined in Example 17.4
   return 0;
}
```

The Overloading Function

The overloading of the operator + for use with objects of the class `complex` is defined by a member function that uses the key word `operator`. Note that this function has one operand, not two. The other operand is the object of which the operator is a member function. Thus in the definition of the function, the first operands are obtained by referring to the variables `real` and `imag` of the class, without reference to a particular object. These operands come from `alpha` when the function is called by `gamma = alpha + beta` in `main()`. Note that we have defined an overloading of only + and not –. Thus a statement of the form

– Needs Definition Also

$$\text{gamma} = \text{alpha} - \text{beta};$$

in `main()` would give a syntax error. A separate function definition for – is required (see Exercise 17.8).

Inline Function

In Example 17.10, we have added `inline` to the definition of the operator + function. Recall (Section 14.2) that this requests that the compiler put the function inline rather than construct a call to the function. (Any function defined within the class itself is implicitly inline.) Some authors suggest that member functions defined outside the class be included in the header file (for example, the file `Complex.h` of Example 17.3), and made `inline`, rather than having them in a

Header versus Implementation Files

separate implementation file (for example, the file of Example 17.4). This keeps the function definitions separate from their prototypes but also allows better efficiency.

As another example of overloading operators, we return to Example 14.5, reproduced here as Example 17.11. In Section 14.3, we used the template of Example 17.11 to generate functions of type `int` and `float`. If the relational operator < had been defined in the previously discussed class `complex`, then we could also use the

max for Complex Objects

`max` function for complex objects. For example, we might define the operator < for complex numbers by comparing the absolute values of the two complex numbers. This is accomplished in Example 17.12 by comparing the squares of the absolute values.

EXAMPLE 17.11
A Template for Maximization Functions

```
template <class T> T max(const T &a, const T &b){
  if(a < b)
    return b;
  else
    return a;
}
```

If the function of Example 17.12 is used in conjunction with the template of Example 17.11 and the class `complex` of, say, Example 17.10, then in the statement `if(a < b)`, the a supplies the arguments `real` and `imag` in Example 17.12. If a < b, the `operator` < function returns 1, which is interpreted as True in the `if` statement; similarly, 0 is interpreted as False. Then we could use the template to generate a `max` function for `complex` objects:

```
complex x, y, z;
//set values of x and y
complex z = max(x, y);
```

Exercise 17.10 asks you to implement the `max()` function. Note that in the template of Example 17.11, the metaparameter `T` would be taken to be `complex`. The template definition would be placed anywhere before the function `main()`, either before or after the definition of the class `complex`.

EXAMPLE 17.12
Overloading the Operator < for Complex Numbers

```
int complex :: operator< (const complex &b){
  if(real * real + imag * imag <
    b.real * b.real + b.imag * b.imag)
    return 1;
  else
    return 0;
}
```

Similarly, if we had a class for rational numbers in which < is defined, we could use max() for objects of this class. In general, max() could be used for any objects for which < is defined, and the template will automatically generate the correct function when needed. On the other hand, the presence of a template does not preclude having an explicitly defined regular function with the same name; for example, max() could also be a function that finds the maximum element of an array.

Some reasonable restrictions apply to overloading operators. For example, we cannot overload the arithmetic operators for the fundamental data types; that is, we cannot redefine + so that 1 + 1 = 0. Also, precedence of the arithmetic operators must be retained. For example, a * b + c must still equal (a * b) + c.

17.3 STREAM CLASSES

cin and cout *Are Class Objects*

We have already used classes and objects in Parts I and II without realizing it: cout and cin are objects that are instances of classes defined in the file iostream.h. In fact, iostream.h defines three basic stream classes: istream for input, ostream for output, and iostream for both. cin is an object of istream, and cout is an object of ostream. Similarly, ofstream and ifstream, used in Section 7.3 for file output and input, are classes defined in the library fstream.h. Objects of all these stream classes are called *streams*; in particular, cin and cout are streams. After our discussion of classes in this chapter, we are now able to understand more completely the use we made in earlier chapters of stream objects.

File Output

In statement (7.13a):

$$ \text{ofstream prt("OUT");} \tag{17.6} $$

we declared an object prt of the class ofstream. Here, prt was an identifier of our choice, and any other identifier could have been used. The argument "OUT" in (17.6) initializes the object prt, so that the file OUT is the current value of the object prt. The effect of this is that a file named OUT is opened and connected to the object prt. Then a subsequent statement like (7.13b):

$$ \text{prt << "My Output Is /n";} $$

sends the message to the file OUT.

Overriding cout

On the other hand, cin and cout are predefined objects that we may use without declaring. cin is automatically connected to the keyboard and cout to the monitor. These are default initializations of these objects but can be overridden by an explicit initialization such as

$$ \text{ofstream cout("OUT");} $$

Now cout is connected to the file named OUT, and subsequent statements with cout will send output to this file, rather than to the monitor. However, it is not recommended that you change the meaning of cout in this way, since it may lead to confusion. Moreover, this cannot be done globally.

In Parts I and II we also used various member functions of the stream classes. For example, in the statement (11.17) in Section 11.3:

```
                     cin.get(c_str, 4);
```

*Use of Member
Functions with* cin
and cout

`get()` is a member function of the class `istream` and is being called using the dot operator notation. Similarly, other member functions were used by the statements:

```
          cin.getline(c_str, 3);
          cout.precision(4);
          cout.width(10);
          cout.setf(ios::fixed);
```

The argument `ios::fixed` of the member function `setf()` is called a *flag*, and `setf` stands for *set flag*. Table 17.1 summarizes some of the more common flags that are used with `setf()`.

open *and* close

Two other member functions that were not previously used are `open()` and `close()`. In place of the statement (17.6), we could write

```
          ofstream prt;
          prt.open("OUT");
```

Here, we first declare an object `prt` of the `ofstream` class and then call the `open()` member function, which connects the stream to the file OUT. Later in the program, we might have the statements

```
          prt.close("OUT");
          prt.open("OUT2");
```

The `close()` member function disconnects the file OUT from the stream, and the next call to `open()` connects the stream to a different file. The same procedure may also be used for opening and closing files for input in conjunction with the class `ifstream`. Closing and opening in this way allows the program to read from, or write to, different files in different parts of the program.

Two other standard objects in `ostream` are `cerr` and `clog`, used primarily to send error messages. For example, we might have the statements

```
          if(q == 0){
            cerr << "Denominator is zero";
            //more code
```

If `cout` were used in this context in place of `cerr`, as has been done throughout the book, the message may be buffered and not sent to the screen immediately.

TABLE 17.1
Flags for `setf()` **Function**

`ios :: scientific`	Floating point numbers printed in scientific notation
`ios :: fixed`	Floating point numbers printed in fixed point notation
`ios :: showpos`	Plus sign printed before positive numbers
`ios :: showpoint`	Decimal point always printed

Using `cerr`, however, the message is not buffered and is sent immediately, which is desirable for error messages. (`clog` also sends to the screen but is buffered.)

A member function that was discussed in Section 7.3 in conjunction with reading from a file is `eof()`, the *end-of-file* function. Still another member function, `fail()`, can be used with either an input or output object to test whether a file has actually been opened. For example, in the code fragment

```
ifstream fin("IN")'
if(fin.fail()){
  cerr << "Failed to open input file";
  exit();
}
```

if the opening of the file `"IN"` fails, `fin.fail()` will return a nonzero value (interpreted as true), and the error message will be displayed. The same construction may be used with output files.

Overloading `>>` and `<<`

We end this section with an example of overloading of the operators `>>` and `<<`. In the class `complex` of Example 17.10, it would be nice if we could write

<code>cin >> beta;</code>

when we wish to input the real and imaginary parts of `beta`. We can achieve this, as well as the corresponding `cout` statement, by overloading the operators `>>` and `<<`. This is illustrated in Example 17.13, in which two objects are read with the statement

<code>cin >> alpha >> beta;</code>

just as if `alpha` and `beta` were ordinary variables.

Description of Example 17.13

We now explain how `>>` and `<<` are overloaded. As discussed in the previous subsection, `istream` and `ostream` are classes defined in the library `iostream.h`. Their names are used as the class names in the function definitions for overloading `>>` and `<<`, just as `complex` was used in Example 17.10 as the class name in the function that overloads `+`. Recall from Section 14.1 that functions may return references; this is accomplished by putting the reference declarator `&` before the function name in the function definition (and the prototype). This is done in Example 17.13, so that the reference declarator `&` after the class names says that the functions will return a reference to objects of type `istream` and `ostream`, respectively. In the function definitions, `in` and `out` are just identifiers, and any other identifiers could be used instead. In the function body for `>>`, the `real` and `imag` parts of the complex object `z` are extracted from the input stream and assigned to `in` by the usual extraction operator `>>`. Then the reference to `in` is returned as the function value. Similar comments apply to the function defining the overloaded operator `<<`.

EXAMPLE 17.13
Overloading >> and <<

```
#include <iostream.h>
class complex{
  private:
    float real; //real part of a complex number
    float imag; //imaginary part of a complex number
  public:
    friend istream &operator >> (istream &,complex &);
    friend ostream &operator << (ostream &,complex &);
};
istream& operator>> (istream &in,complex &z){ //>> function
  cout << "what are real and imag?\n";
  in >> z.real >> z.imag;
  return in;
}
ostream& operator<< (ostream &out, complex &z){ //<< function
  out << "real=" << z.real << " imag=" << z.imag << endl;
  return out;
}
int main(){
  complex alpha, beta; //declare two objects
  cout << "read alpha and beta\n";
  cin >> alpha >> beta; //read two complex numbers
  cout << alpha << beta << endl; //display two complex numbers
  return 0;
}
```

A Wrong Approach

Note that the overloading of >> and << is done in a quite different manner from that of the operator +. Had we followed the same form as for +, we would have used a function prototype like

```
complex operator>> (complex); //incorrect
```

But this will not work. We must use the stream classes `istream` and `ostream`, and the return type of the functions must be of one of these classes.

A Different Approach Using Manipulator Functions

The second point to note is that the overloading functions are defined as `friend` functions. This is necessary to allow the dummy object z in the functions to have access to the private variables `real` and `imag`. Alternatively, we might try to define the overloading functions as member functions of the class, but this will not allow us to use the usual `cin >> gamma` form of the `cin` statement. To achieve this, the function must be defined outside the class, but rather than using `friend` functions, we could use manipulator member functions, such as were used in Example 17.1. That is, we could define the manipulator function

```
void complex :: set(float c, float d){
  real = c;
  imag = d;
}
```

Then the overloading function for >> could be defined by

```
istream& operator>> (istream &in, complex &z){
   float real, imag;
   in >> real >> imag;
   z.set(real,imag);
   return in;
}
```

A similar change would be done for the function overloading <<. The prototypes for both overloading functions would now be removed from the class definition. It is left to Exercise 17.9 to write a complete program using this approach.

Reference Returns
More Efficient

The return of a reference is necessary in the overloaded functions for << and >> in Example 17.13, and it may also be desirable in other situations. If the return of a function is not a reference, the return value will be a temporary object, whereas if the return is a reference, no temporary object is created. For the simple classes of this chapter, there will be no great savings in efficiency by using reference returns, but for more complicated classes the benefit may be worthwhile.

Modularization and Software Reuse

We have previously discussed how functions can be used to modularize a program and also contribute to software reuse. Classes can help accomplish the same goals. The classes iostream, istream, and ostream play a role similar to that of library functions: They are available for use in any C++ program. Similarly, user-defined classes can play the same role as used-defined functions.

We have used complex numbers as the main example in this chapter to illustrate various aspects of defining and using classes. In particular, we discussed how:

1. Complex numbers can be declared and initialized.

2. The usual operations of +,-, *, and / can be used with complex numbers.

3. The insertion and extraction operators << and >> may be used with complex numbers.

4. Arrays of complex numbers may be declared.

Thus once a suitable class complex has been defined, complex numbers may be used in much the same way as the fundamental type float. Even this relatively simple class for complex numbers is a package that contains several functions as well as the data themselves.

After sufficient testing, such a class could be used by you or others in any number of programs. Note the advantage of information hiding here: Neither you nor another user of this class need to worry about the identifiers that are used for the real and imaginary parts of a complex number in the class definitions, and are free to use the same identifiers in a program that uses the class.

MAIN POINTS OF CHAPTER 17

- A class is a user-defined data type, called an abstract data type, consisting of variables, and functions that operate on these variables. An object is a particular instance of a class.
- Classes are valuable for information hiding, for software reuse, and for fault containment.
- Classes consist of private and public member variables and functions. Only public members are directly available outside the class. The private members provide information hiding.
- Member functions of a class are accessed by the access operator " . ".
- A simple example of a class consists of complex numbers and functions that do computation with complex numbers.
- Objects may be function parameters and return values.
- Friend functions are nonmember functions that may access the private variables of a class.
- `cin` and `cout` are objects of the stream classes `istream` and `ostream`.
- Operators such as + and > may be overloaded so as to be used on objects. The insertion and extraction operators << and >> may also be overloaded, but this should be done in a different way.
- A constructor is used to initialize objects when they are declared. The constructor may contain parameters that set initial values of one or more of the private variables of the object to the values given by the parameters. Default arguments may also be used.
- A default constructor is used automatically by the compiler when an object is declared and no values of private variables are specified. A default constructor does not need to appear explicitly in the class definition unless an explicit constructor is also used. A default constructor has no parameters and usually no body, although a body may be supplied to give specific initial values to some of the private variables.

EXERCISES

17.1   Run the program of Example 17.2 [including the class of Example 17.1, which you insert before `main()`] for various values of `alpha`.

17.2   The quotient of two complex numbers `a + ib` and `c + id` is defined by
$$\frac{a+ib}{c+id} = \frac{(a+ib)(c-id)}{(c+id)(c-id)} = \frac{ac+bd+i(bc-ad)}{c^2+d^2}$$

Add to the class of Example 17.3 a member function that carries out complex division.

17.3   Run the program of Examples 17.3–17.5 for various values of `beta`. Run the program again with the default constructor of Example 17.3 omitted and observe the syntax error.

17.4 Modify the program `main()` in Example 17.5 to declare and use arrays of complex numbers.

17.5 Rewrite the function `mult()` of Example 17.4 as an auxiliary function corresponding to the function `add()` in Example 17.7.

17.6 Add a friend function to do addition of complex numbers to the program of Example 17.8, as discussed in the text.

17.7 Add to the class of Example 17.3 a member function `abs_value()`, as defined in Example 17.8, but put only the prototype in the class and then put the function definition after the class.

17.8 Add overloaded operators for * and – in Example 17.10 so as to do multiplication and subtraction of two complex numbers.

17.9 Rewrite the program of Example 17.13 using manipulator member functions along the lines discussed in the text.

17.10 Add the operator < function of Example 17.12 to the `complex` class of Example 17.10 and use the template of Example 17.11 to generate a `max` function for complex objects, as discussed in the text. Write a function `main()` that utilizes this `max` function.

18

ARRAY CLASSES AND DYNAMIC MEMORY

The examples of classes so far have demonstrated rather simple uses of the object-oriented paradigm. The ability to define new types, as well as operations on those types, is a powerful feature. In this chapter, we present a number of array classes that may be used for vector and matrix operations.

18.1 A VECTOR CLASS

As a first example, we give in Example 18.2 an array class in which the private member is an array of length 3. This class has member functions that form the sum of two vectors and the dot product of two vectors. The operator + is overloaded to perform vector additions but the dot product does not use overloading since a.dot(b) is quite natural notation. However, Exercise 18.1 asks you to overload * for the dot product. We also overload >> and << so as to be able to read and display vectors using these operators. Note that the class in Example 18.2 does not have any explicit constructors since no objects are initialized in the function `main()` of Example 18.1.

EXAMPLE 18.1
Using the Vector Class

```
int main(){
   vector a,b,c;   //declare three objects
   cin >> a >> b; //read vectors a and b
   c = a + b;      //add a and b
   cout << c;      //display c
   float dp = a.dot(b);//dot product of a and b
   cout << "dot = " << dp << endl;
   return 0;
}
```

EXAMPLE 18.2
A Vector Class

```
#include <iostream.h>
class vector{
  private:
    float vec[3];
  public:
    vector operator+ (const vector &); //operator prototype
    float  dot(const vector &);//prototype for dot product
    friend istream& operator >> (istream &, vector &);
    friend ostream& operator << (ostream &, vector &);
};
istream& operator>> (istream &vecin, vector &v){
  cout << "what is the vector?\n";
  for(int i = 0; i < 3; i++)
    vecin >> v.vec[i];
  return vecin;
}
ostream& operator<< (ostream &vecout, vector &v){
  for(int i = 0; i < 3; i++)
    vecout << "element " << i << " of vector is "
           << v.vec[i] << endl;
  return vecout;
}
vector vector :: operator+ (const vector &u){ //overloaded +
  vector v;
  for(int i = 0; i < 3; i++)
    v.vec[i] = vec[i] + u.vec[i];
  return v;
}
float vector :: dot(const vector &u){ //dot product
  float value = 0;
  for (int i = 0; i < 3; i++)
    value = value + vec[i] * u.vec[i];
  return value;
}
```

18.2 DYNAMIC MEMORY ALLOCATION

We may use dynamic memory allocation (see Chapter 16) of objects. For example, in the context of the class `complex` of the previous chapter the statements

```
int n;
cin >> n;
complex *alpha = new complex[n];
```

define `alpha` to be a pointer to n complex objects, so that `alpha` is an array of objects with memory dynamically allocated. The default constructor of the class `complex` initializes each object separately to zero. Another example is

```
complex *beta = new complex(1.0, 2.0);
```

Here, storage for a single complex object is allocated, and the constructor initializes this storage to the indicated values.

Our use of dynamic memory allocation in the remainder of this chapter will be for array classes. The class in Example 18.2 has a serious limitation: It is restricted to vectors of length 3. Ideally, the class should be able to handle vectors of any length. We next give a class that does this, by using dynamic memory allocation. Following the style of Examples 17.3–17.5, a header file is shown in Example 18.3 followed by an implementation file in Example 18.4. In Examples 18.3 and 18.4, member functions read() and print() are defined. Exercise 18.3 asks you to replace these functions to overload >> and <<, analogous to what was done in Example 18.1.

EXAMPLE 18.3
Header File for the Vector Class

```
#include <iostream.h>
class vector{
private:
    int veclength;
    float *vec;
public:
    vector(); //default constructor prototype
    vector(int); //constructor prototype
    ~vector(); //destructor prototype
    vector(const vector &);//copy constructor prototype
    vector operator+ (const vector&); //operator prototype
    float& operator[ ] (int); //[ ] prototype
    vector& operator= (const vector&); // = prototype
    void read();
    void print();
};
```

The dynamic memory allocation is achieved by defining a pointer, vec, and then the constructor allocates storage by the new operator. This is the same procedure used in Example 16.2, except that here the allocation of storage is done by the constructor. Note that the desired length of the vector is the argument of the constructor, which assigns this value to the private variable veclength. Thus the statement

Constructor Does the Memory Allocation

```
vector a(n);
```

in main() in Example 18.5 declares a vector of length n. In the implementation of the vector operations, the length of the vectors must be known and is set by statements of the form int n = u.veclength. Note that Example 18.4 contains a default constructor that sets veclength and vec to zero; the usual default constructor does not necessarily set these parameters to zero. Although not needed in this simple example, this type of default constructor is sometimes necessary when working with dynamic memory.

EXAMPLE 18.4
Implementation File for the Vector Class

```
vector :: vector(){ //default constructor
   veclength = 0;
   vec = 0;
}
vector :: vector(int n){ //constructor
   veclength = n;
   vec = new float[n];
}
vector :: ~vector(){ //destructor
   if(vec != 0) delete [ ]vec;
}
vector :: vector(const vector &v){ //copy constructor
   veclength = v.veclength;
   vec = new float[veclength];
   for(int i = 0; i < veclength; i++)
      vec[i] = v.vec[i];
}
vector vector :: operator+ (const vector &u){ //overloaded +
   int n = u.veclength;
   vector v(n);
   for(int i = 0; i < n; i++)
      v.vec[i] = vec[i] + u.vec[i];
   return v;
}
float& vector::operator[ ](int i){ // [ ] definition
   return vec[i];
}
void vector :: read(){
   for(int i = 0; i < veclength; i++){
      cout << "array element = ?\n";
      cin >> vec[i];
   }
}
void vector :: print(){
   for(int i = 0; i < veclength; i++)
      cout << "element " << (i + 1) << " = " << vec[i] << endl;
}
```

Example 18.3 has another member function of a different type:

~vector() {if(vec != 0) delete [ ]vec;}

Destructor to
Delete Memory

The ~ indicates that this is a *destructor function*. The purpose of a destructor is to delete memory allocated by the new operator, when this memory is no longer needed. For example, when the function for + in Example 18.4 has finished executing, the memory for the local object v defined in the function body is deleted. More generally, a destructor is called whenever an object goes out of scope or the program terminates. The destructor is never called explicitly but is invoked automatically at the appropriate times. The destructor does not delete the object on which it is

EXAMPLE 18.5
Using the Vector Class with Dynamic Memory Allocation

```
int main(){
   int n;
   cout << "n = ?\n";
   cin >> n;
   vector a(n); //declare an object
   a.read(); //read vector a
   for(int i = 0; i < n; i++) //display a
     cout << a[i] << endl; //this statement needs the [] function
   vector b = a; //this statement uses the copy constructor
   b.print();
   vector c(n);
   c = a + b; //this statement needs the operator = function
   c.print();
   return 0;
}
```

invoked; rather it deletes any memory allocated within that object. Because the default constructor sets `vec` and `veclength` to zero, we know that if they are nonzero, the other constructor has set their values. Then in the destructor, we test to see if `vec` is zero, and if not we delete the allocated memory. Note that we do not need to also set `vec` to zero to prevent a dangling pointer, because the memory containing the vector object is to be reclaimed.

Copy Constructors

The class `vector` of Examples 18.3 and 18.4 also contains other constructors that are necessary when using dynamic memory allocation. The first is a *copy* constructor,

Copy Constructor   which is used in statements such as

```
vector b = a; //This statement uses the copy constructor
```

This statement, which appears in `main()` in Example 18.5, declares an object `b` and sets it equal to `a`. [Here, it would be incorrect to write `b(n) = a` or `b(n) = a(n)` since `a` has been previously declared with length n.] Usually, for such a statement the compiler supplies an adequate copy constructor, and no explicit constructor is required. In particular, if `a` had been defined as an object of the vector class of Example 18.1, the default copy constructor would copy the array `a` element by element to `b`. However, when `a` is defined by means of dynamic storage allocation, as in Examples 18.3 and 18.4, the default copy constructor is not adequate because it will copy only the members in an object, in this case `vec` and `veclength`. This is called a *shallow* copy. Both `a.vec` and `b.vec` now point to the same storage space and would just be aliases; consequently, if the data in `a` are

Deep versus   changed, `b` will also be changed, which is unacceptable. What is needed is a copy
Shallow Copy   constructor that will allocate storage for the new object as well as copy the members

of the object. This is called a *deep* copy, and is illustrated in the copy constructor in Example 18.4, in which veclength is copied, storage is allocated for vec, and the elements of the array are copied into this new storage. Copy constructors are also invoked when an object is passed by value to a function or is a return value of a function.

Indices and operator[ ]

The vector class of Examples 18.3 and 18.4 contains another function that needs discussion. Recall that to access the private members of a class from outside the class one normally needs some kind of accessor function. In particular, if we wished to access elements of the vector a, we could have an accessor function, say element(), so that a.element(i) would give the ith element of the array. However, this is clumsy, and it would be much better to use the usual array notation a[i]. But to be able to do this, we must first define the operator[ ]. Although [ ] may not seem like an operator, consider x[i], where x is some one-dimensional array. Then, as discussed in Chapter 15, x[i] is equivalent to *(x + i), where x is now considered to be a pointer to the first position of the array and [ ] may be viewed as an operator that effects the address translation. The function

$$\text{float\& operator[ ](int i){return vec[i];}} \qquad (18.1)$$

Defining
operator[]
in Examples 18.3 and 18.4 defines the operator[ ] and allows direct addressing of the elements of an array, as in the cout statement in Example 18.5. Without this function for [ ] the cout statement would be illegal and would give a syntax error on compilation. Note that the operator[ ] function returns a reference, which is interpreted as the address of a[i] in the cout statement. The function would be incorrect without the reference declarator.

As we have mentioned many times, the usual mathematical notation for vector elements, $x_1, \ldots, x_n$, does not match the C++ convention of array indices beginning with 0. However, we may overload the operator [ ] so that array indices do, indeed, begin with 1. To do this, we would replace in Examples 18.3 and 18.4 the function
Shifting Indices   (18.1) by

$$\text{float\& operator[ ](int i){return vec[i - 1];}} \qquad (18.2)$$

Then in the function main() of Example 18.5 we could write the cout loop as

```
for(int i = 1; i <= n; i++) //display a
cout << a[i] << endl;                                    (18.3)
```

Here, the loop indices run from 1 to n. Note, however, that the indices in the member functions of Example 18.4 are not to be changed since these use the private variable vec. It is left to Exercise 18.4 to make the changes (18.2) and (18.3) in Examples 18.3–18.5. We caution that since this kind of indexing is not typical in C++, it should be well documented if it is used.

Bounds Checking

Another thing we might like to add to the `vector` class is *bounds checking*. Recall that C++ does not check to see if an array index is an acceptable value. But we can build this check into our class definition if we like. It could be added to the definition of the operator [ ] function, as shown in Example 18.6. Exercise 18.5 asks you to modify this function so it can be used in conjunction with the shifted index definition (18.2).

EXAMPLE 18.6
An `operator[ ]` Function with Bounds Checking

```
float& vector :: operator[ ](int i){
  if(i < 0 ||i >= veclength)
    cerr << "Array index out of bounds";
  return vec[i];
}
```

Assignment and `operator=`

If `c` and `b` are two objects of the `complex` class, it would be nice to be able to write an assignment statement such as

c = b; //ok for complex objects

This is in fact valid. Similarly, if `a` and `b` are two objects of the vector class of Example 18.1, we could write the assignment

a = b; //ok for vectors, if no dynamic memory

However, if `a` and `b` are objects of the vector class of Example 18.3, in which memory has been allocated dynamically, an assignment statement such as the above can lead to difficulty.

In an assignment statement, the compiler automatically invokes a copy operation, essentially the same as that used in the default copy constructor. This operation copies each member of the object on the right side of = to the corresponding member on the left side. This is again a shallow copy and gives rise to the same problem as with the default copy constructor if the objects have been allocated storage dynamically. Thus, if `a` and `b` are two objects of the vector class of Example 18.3, the assignment a = b; would copy the pointer `vec` of the `b` object to that of the `a` object. The previous pointer `vec` in `a` is destroyed and the storage to which it pointed is now lost. This is one classic cause of a memory leak, and is depicted in Figure 18.1.

Memory Leak

Figure 18.1 Memory Leak from Assignment

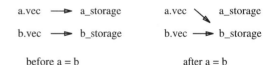

before a = b   after a = b

Aliasing

Moreover, a and b are now aliases and any change of one will also change the other. This problem does not occur in the function `main()` of Example 18.5. However, suppose (see Exercise 18.8) we added at the end of `main()` the statements

```
a = c;
c[0] = 1;
```

The first statement makes c an alias of a so that the second statement changes a[0] as well as c[0].

To circumvent this aliasing problem, as well as the memory leak, we need to add to the vector class an explicit function for the operator =. Such a function is shown in Example 18.7.

EXAMPLE 18.7
An Operator = Function for the Vector Class

```
//vector& operator= (const vector&); //prototype
vector& vector :: operator= (const vector &v){ //operator =
   if(this != &v){ //check for statements like a = a
      if(vec != 0) delete [ ]vec;
      veclength = v.veclength; //set veclength
      vec = new float[veclength]; //allocate storage
      for(int i = 0; i < veclength; i++) //copy the elements
         vec[i] = v.vec[i];
   }
   return *this;
}
```

The this *Pointer*

In the function of Example 18.7, we delete the previous storage allocated to the object on the left side of = and then allocate new storage; this prevents a memory leak as well as the aliasing problem. We then copy the elements of the vector. In the implementation of this function we use `this`, which is a special predefined pointer that points to the calling object, that is, to the object on the left side of the assignment operator. This construction is necessary if we should have an assignment statement like a = a; in this case, `*this` is returned and nothing else is done. Otherwise, we still return `*this`, since `this` is now the pointer to the new left-hand-side object. We note that `this` is a keyword and cannot be explicitly declared.

The `operator=` function of Example 18.7 should be added to the vector class. The prototype shown in Example 18.7 was already put in the header file of Example 18.3 in anticipation of this function. The implementation of the function should be added to Example 18.4 (see Exercise 18.8).

In order to avoid unforeseen problems, a general rule of thumb is this: If a class has a copy constructor or an `operator=` function or a destructor, it should have all of these.

Class Templates

In Chapter 14 we discussed templates for generating functions. A class template may be used in much the same way to generate classes, as illustrated in Example

18.8. Here, T is a type metaparameter and n is an int type metaparameter. Note that, just as with any class definition, a semicolon following the closing brace is necessary.

The private variable of this template is an array whose type and size are both determined when the class is instantiated. There is also a public member function sum() whose return type is determined when the class is instantiated. For example, the statement

$$\text{vector <float, 5> FloatVector;} \qquad (18.4)$$

would instantiate an object FloatVector in which the private variable vec is an array of type float and length 5 and whose public member function sum() has return type float. Here the metaparameter T in Example 18.8 is replaced by float in the generation of the class. We could use this template by a function main() as shown in Example 18.9. (Note that the operator[ ] function is necessary in the class template in order to use FloatVector[i] in main().)

EXAMPLE 18.8
A Class Template

```
template <class T, int n> class vector{
  private:
    T vec[n];
  public:
    vector(){}; //constructor
    T &operator[ ](int i){return vec[i];}
    T sum(){
      T value = 0;
      for(int i = 0; i < n; i++)
          value = value + vec[i];
      return value;
    }
};
```

EXAMPLE 18.9
Using the Class Template vector

```
#include <iostream.h>
//class template here
int main(){
  vector <float, 5> FloatVector;
  cout << "What is FloatVector? \n";
  for(int i = 0; i < 5; i++)
    cin >> FloatVector [i];
  cout << FloatVector.sum() << endl;
  return 0;
}
```

On the other hand, the statement

`vector <int, 20> IntVector;`

would instantiate an object `IntVector` whose private variable `vec` is an array of type `int` and length 20. Here, the metaparameter `T` is replaced by `int`. Similarly,

`vector <complex, 10> ComplexVector;`

would instantiate an object corresponding to an array of 10 `complex` objects. Thus the class template allows us easily to generate arrays of different types and lengths. Note, however, that the template of Example 18.8 can only generate classes with fixed length vectors; that is, the value of the parameter `n` in the template definition must be known when statements such as (18.4) are compiled. Exercise 18.11 asks you to define a template that will generate a class that dynamically allocates storage for vectors.

18.3 A MATRIX CLASS

We next construct a matrix class that is defined in terms of the `vector` class of Examples 18.3 and 18.4. We assume that the `operator=` function of Example 18.7 has been added to the `vector` class as well as the function shown in Example 18.10. The reason for the latter will be discussed shortly.

EXAMPLE 18.10
Function to be Added to Vector Class

```
void allot(int); //prototype
void vector :: allot(int n){
  if(vec != 0) delete []vec;
  veclength = n;
  vec = new float[n];
}
```

In the definition of the class `matrix` in Example 18.11, `A` is a pointer to vectors. Defining `A` as a pointer to vectors allows us to use the usual array notation `A[i][j]`, just as in Example 16.5, where `A` was defined as a pointer to pointers. The reason is that since `A` is a pointer to vectors, `A[i]` is a vector reference so that `A[i][j] = (A[i])[j]`. However, we need to define the `operator[ ]` to return `A[i]` in order to achieve this.

EXAMPLE 18.11
A Matrix Class with Dynamic Memory Allocation

```
//This class uses the vector class of Examples 18.3 and 18.4
//augumented by the member functions of Examples 18.7 and 18.10
#include <iostream.h>
class matrix{
  private:
    int n;
    vector *A;
  public:
    matrix(){A = 0;}//default constructor
    matrix(int N);//constructor prototype
    ~matrix() {if(A != 0) delete [ ]A;}//destructor
    vector &operator [ ] (int i){return A[i];}
    matrix(const matrix &);//copy constructor prototype
    matrix& operator =(const matrix &);//prototype
    void read();
};
matrix ::  matrix(int N){//constructor
  n = N;
  A = new vector[N]; //allocate array of vectors
  for(int i = 0; i < N; i++) //allocate storage
    A[i].allot(N);
}
matrix :: matrix(const matrix &B){//copy constructor
  int N = B.n;
  A = new vector[N];          //allocate
  for(int j = 0; j < N; j++)   //new storage
    A[j].allot(N);
  for(int i = 0; i < N; i++)   //copy
    for(int j = 0; j < N; j++) //the
      A[i][j] = B[i][j];       //matrix
}
matrix& matrix :: operator=(const matrix &B){//operator=
  if(this != &B){
    if(A != 0) delete [ ]A;//delete storage
    int N = B.n;
    A = new vector[N];          //allocate
    for(int j = 0; j < N; j++)   //new
      A[j].allot(N);            //storage
    for(int i = 0; i < N; i++)
      for(int j = 0; j < N; j++) //copy the
        A[i][j] = B[i][j];       //matrix
  }
  return *this;
}
void matrix ::  read(){
  cout << "input " << n << " squared values for matrix\n";
  for(int i = 0;i < n;i++)
    for(int j = 0; j < n; j++)
      cin >> A[i][j];
}
```

EXAMPLE 18.12
Using the Matrix Class

```
int main(){
  int N;
  cout << "N = ?\n";
  cin >> N;
  matrix C();//declare a matrix
  C.read(); //read the matrix
  for(int i = 0; i < N; i++) //print the matrix
    for(int j = 0; j < N; j++)
      cout <<  C[i][j] << endl;
  return 0;
}
```

In the constructor for the matrix class, the new operator allocates storage for N pointers to vectors; we can consider these as pointers to the rows of a matrix. Then storage is allocated to the ith row by the statement A[i].allot(N), where allot() is the member function given in Example 18.10. This function does exactly the same thing as the constructor in the vector class, but it is a member function. We cannot call the constructor function from the matrix class in the same way, because it is not possible to pass parameters to a constructor in such a way as to allocate storage for the rows of A.

A similar situation exists with regard to deleting memory. The destructor is always called automatically and cannot be called explicitly. Thus, if you wish to delete the memory assigned to an object, you need a member function to do this. Such a member function could be given by

Deleting Memory for the vector *Class*

```
void vector :: deallot(){
  delete [ ] vec;
  vec = 0;
  veclength = 0;
}
```

When one class uses another class, as the matrix class in Example 18.11 uses the vector class, the class that is used must of course be available. The most straightforward way to make it available is simply to add the definition of the vector class to the matrix file. However, it can also be made available conveniently by the extern statement, as illustrated below:

The extern *Statement*

```
#include <iostream.h>
extern vector A;
class matrix{
```

Here, the statement extern vector A; has been added before the definition of the matrix class, and informs the compiler that the definition of the vector class is available in another file.

Global constants or functions that have been defined in another file may also be made available through the `extern` statement, as illustrated by:

```
extern float PI;
extern int func(int i);
```

MAIN POINTS OF CHAPTER 18

- Array classes for vectors and matrices may be defined so that memory is dynamically allocated when an array object is declared.

- When memory has been allocated dynamically (and in a few other special situations), a copy constructor, an assignment constructor, and a destructor to delete memory are needed. If one of these is present, usually all should be present.

- A copy constructor is used to make a deep copy of an object; that is, memory is allocated for the new object, and the member variables copied to this new memory. Copy constructors are used to copy objects in declarations, and when objects are used as arguments or return values of functions.

- An assignment constructor (an `operator=` function) is used to prevent aliasing and memory leaks when assignments are made with dynamically allocated objects. The `operator=` function deletes the current memory of the left-hand-side object, allocates new memory, and copies the right-hand-side object. The `operator=` function is usually implemented using the pointer `this`, in order to be able to handle situations such as `a = a`.

- Templates may be used to generate classes automatically depending on certain operators.

EXERCISES	

18.1 Modify the program of Examples 18.1 and 18.2 by using an overloaded operator `*` so that `a * b` is the dot product of two vectors.

18.2 Add a dot product member function to the class of Examples 18.3 and 18.4.

18.3 Replace the functions `read()` and `print()` of Examples 18.3 and 18.4 by functions to overload `>>` and `<<`, as was done in Example 18.1.

18.4 Replace the `operator[ ]` function in Examples 18.3 and 18.4 by the function (18.2). Test the new class by changing the `cout` loop in Example 18.5 to (18.3).

18.5 Add bounds checking, as done in Example 18.6, to the shifted index definition (18.2) of the `operator[ ]` function.

18.6 The norm (length) of a vector **a** with components $a_0, \ldots, a_{n-1}$ is defined as

$$\left(\sum_{j=0}^{n-1} a_j^2 \right)^{1/2}$$

Add a member function for the norm to the class in Examples 18.3 and 18.4.

18.7   Give an overloading of the operator $*$ that defines a member function of the vector class to carry out the vector–scalar multiplication $v * scale$, where v is a vector and scale is a float. Can you define the function so that the operands can be reversed? That is, can you compute scale $*$ v?

18.8   Add the statements a = c; c[0] = 1; at the end of the function main() of Example 18.5 and observe that a[0] also changes when c[0] is changed. Now add the operator= function of Example 18.7 to the vector class of Examples 18.3 and 18.4 and observe that the above two assignment statements do not change a[0].

18.9   Define an overloading of operator< for objects of the vector class of Examples 18.3 and 18.4 by comparing norms (see Exercise 18.6) of the vectors. Then use this operator< function with the template of Example 17.11 to generate a max function for vector objects. Write a function main() to test the max function.

18.10 Use the template of Example 18.8 in conjunction with a function main() that will declare an int vector of length 20 and a float vector of length 10.

18.11 Modify the template of Example 18.8 so that it will handle vectors of arbitrary length n. You should take the following steps:

   a.   Eliminate the argument n in the template argument list and add a private variable n.

   b.   Add a constructor, as in Example 18.4, that will dynamically allocate storage.

   c.   Declare a vector in main() by a statement such as

```
vector <float> x(n);
```

   Here, n should first be read by a cin statement.

   Test your template by declaring vectors of both type int and type float in main(). Can you add a destructor, copy constructor, operator= function, and operator[ ] function to the template class?

18.12 Write a function that does matrix-vector multiplication. The arguments of the function should include a matrix that is an object of the matrix class of Example 18.11 and a vector that is an object of the vector class of Examples 18.3 and 18.4.

18.13 Define a class for $m \times n$ matrices independently of a vector class. It should contain int m,n and float *A as private variables and allocate storage by a constructor of the form

```
matrix(int M, int N){A = new float [M * N]; m = M; n = N;}
```

Add the other necessary copy and assignment constructors and test the class. Discuss the advantages and disadvantages of this approach to a matrix class, as compared with that of Example 18.11.

19

INHERITANCE

There are two main ways that classes may use other classes. The first is that members of a class may be members of another class. For example, the `matrix` class of Example 18.11 had a private member of the `vector` class. In this case, we say that the `matrix` class is a *client* of the class `vector` or that it *contains* `vector`. The second way classes may use other classes is a very powerful feature called *inheritance*. Inheritance allows us to obtain new classes, called *derived classes*, from existing or *base* classes. Inheritance sometimes avoids the necessity of rewriting the base class in order to makes changes to it. Indeed, the derived class may sometimes be produced without even having the source code of the base class. We note that the classes `istream` and `ostream` discussed in Chapter 17 are derived from a class called `ios`. `iostream`, in turn, is derived from both `istream` and `ostream`, and still other classes are derived from `iostream`.

Inheritance is an extensive and difficult topic. The purpose of this chapter is simply an introduction to the concepts of inheritance as well as polymorphism and virtual functions.

19.1 DERIVED CLASSES

The `is a`
Principle

The most important principle guiding the construction of derived classes is the "*is a*" relationship:

Every object of the derived class is also an object of the base class.

We will illustrate this principle with our first example of a derived class.

For many problems in scientific computing, complex numbers whose real part is positive play a special role. (In some cases, it is complex numbers whose real part is negative, rather than positive.) In Example 19.1 we give a derived class, `ComplexPos`, for complex numbers with positive real parts. This class is derived from the `complex` class of Example 17.10, to which we assume that a member

EXAMPLE 19.1
A Derived Class

```cpp
#include <iostream.h>
class complex{//This is the interface for the class of Example 17.10
  protected: //The following are available to derived classes
    float real;
    float imag;
  public:
    complex() {} //constructors
    complex(float re, float im){real = re; imag = im;}
    void read();
    void print();
    complex operator + (const complex &);
    complex operator * (const complex &); //from Exercise 17.8
};
//This defines a derived class based on the class complex
class ComplexPos : public complex{
  public:
    ComplexPos() : complex() {} //Default constructor
    ComplexPos(const complex &a) : complex(a){}//constructor
    ComplexPos(float,float);//constructor prototype
    ComplexPos &operator = (const ComplexPos&);//operator = prototype
    void read(); //new read function prototype
  private:
    void error_print(char [ ]); //error message prototype
};
ComplexPos :: ComplexPos(float re,float im){//constructor
  real = re;
  imag = im;
  if(re <= 0) error_print("Constructor");
}
ComplexPos& ComplexPos :: operator = (const ComplexPos &beta){
  real = beta.real;
  imag = beta.imag;
  if(beta.real <= 0) error_print("operator =");
  return *this;
}
void ComplexPos :: read(){
  cout << "enter real and imag parts\n";
  cin >> real >> imag;
  if(real <= 0) error_print("Read Function");
}
void ComplexPos :: error_print(char c[ ]){
  cout << "Message from " << c << ":\n";
  cout << "this number does not have positive real part\n";
  cout << "real = " << real << "  imag = " << imag << endl;
}
int main(){
  ComplexPos alpha(-1,8), beta, gamma, delta;
  beta.read(); //read a complex number
  gamma = alpha + beta;
  gamma.print(); //display the sum
  delta = alpha * beta;
  delta.print(); //display the product
  return 0;
}
```

function for overloaded multiplication has been added (Exercise 17.8). The interface of this base class is also shown in Example 19.1. Clearly, every complex number with positive real part is also a complex number so the "is a" relationship between the derived and base classes is satisfied.

In main(), we declare alpha, beta, gamma, and delta to be objects of the derived class ComplexPos. We can now use the overloaded operators * and + on these objects, as well as the read() and print() member functions of the base class. One advantage of the use of derived classes is that we can extend the base class with essentially no changes to it. In the simple example of Example 19.1, this advantage is perhaps not evident. However, if the base class is large, complicated, and perhaps written by somebody else, it is much safer to use it as is and extend it by a derived class, rather than try to modify the base class.

Protected Members

We next discuss a number of aspects of the program in Example 19.1. First, in the base class we have declared the variables real and imag to be protected, a designation we have not previously used. In the classes we defined in the previous chapters, all variables were declared to be private. However, if they are private, they can be accessed only by member functions of the class in which they are declared and not even by member functions of a derived class. We could circumvent this problem by declaring real and imag to be public. But this would allow access to them by *any* other program, which is what we tried to avoid by the private designation. But there is a third possibility: protected. Then, real and imag are available to a derived class but are still private to other programs; this is just what we desire. This is the one change we must make to the base class in order to derive the ComplexPos class from it. Note that protected members are used only in conjunction with derived classes, but protected variables may also be accessed by friend functions.

Derived Class Definition

Following the base class complex, we define the derived class, ComplexPos. The first line of this class definition is

```
class ComplexPos : public complex{
```

which gives the name of the derived class followed by the name of the base class. The derived class contains public member functions for three constructors, as well as operator= and read. We consider first the constructor with prototype

$$\text{ComplexPos(float, float);} \tag{19.1}$$

Constructor

This constructor is used in main() for the initialization of alpha. The implementation of this constructor contains a test to ensure that the real part of the number is indeed positive. If not, it calls the member function error_print(), which displays a message that the number does not have positive real part. The operator= and the read() functions will also display similar messages. Therefore, the error_print() function accepts as a parameter a string indicating from which function the error

Error Message

message is being sent. A typical error message is:

```
Message from Constructor:
This number does not have positive real part
real = -1  imag = 8
```

In error messages from the other two functions, the word `Constructor` will be replaced by `operator=` or `Read Function`. Except for `error_print()`, all the member functions of `complex_pos` are `public`. `error_print()` is meant to be called only by the other member functions, and so is made `private`; it is thus not available to users of the class.

In the previous section we stated that an `operator=` function was needed in conjunction with dynamic memory allocation. Since there is no dynamic memory allocation in the class `ComplexPos`, there would be no need for an `operator=` function if we just wished to carry out assignments. However, we wish the assignment function to test whether a complex number with positive real part is being assigned, and we need to build this test into an `operator=` function.

Assignment

The function `read()` in the `ComplexPos` class has the same name as the corresponding function in `complex`. There is no conflict here, however, since the name in the derived class *dominates*, and it is this function that is used. Access to the function `read()` in the `complex` class may be achieved by means of the scope resolution operator and base class name. For example, the statement

Domination

`beta.complex::read();`

would access the `read()` function from the `complex` class.

There are two other constructors in the `ComplexPos` class that play different roles than the constructor with prototype (19.1). The constructor

Other Constructors

`ComplexPos(const complex &a):complex(a){}`   (19.2)

is a special case of a constructor with the general form

`DerivedClassName(arguments) : BaseClassName(arguments) {code};`

in which a derived class constructor is defined in terms of a base class constructor. The particular constructor (19.2) has only the argument `complex a` and no function body. The purpose of this constructor is to call the base class constructor so as to make objects of `ComplexPos` also objects of `complex` for the purpose of using the member functions in the class `complex`. Without the constructor (19.2), the statements

`gamma = alpha + beta; delta = alpha * beta;`

in `main()` will give syntax errors on compilation (see Exercise 19.1). This is because `+` and `*` are member functions of the base class and can be referenced only by objects of type `complex`. The default constructor is just the special case (19.2) in which there are no parameters.

Private, Protected, and Multiple Inheritance

In the definition of the derived class of Example 19.1, the base class was declared as `public`. It is also possible to declare the base class as private (in which case public member functions of the base class become private members of the derived class and are not accessible outside the derived class) or protected (in which case public member functions of the base class become protected members of the derived class). Private and protected inheritance are used rather infrequently, and we shall not consider them further.

A derived class may be the base class for still another derived class. Hence, it is possible to construct a hierarchy of derived classes. Classes may also be derived from two or more base classes, which is called *multiple inheritance*. The syntax for the declaration is

```
class DerivedClass : public BaseClass1, public BaseClass2
```

In this case, the derived class would inherit the properties of both base classes. A full discussion of multiple inheritance is beyond the scope of this book.

19.2 POLYMORPHISM AND VIRTUAL FUNCTIONS

Suppose that we have a class that computes areas of rectangles and a derived class that computes areas of squares, as shown in Example 19.2. (Clearly, the "is a" relationship is satisfied since a square is a rectangle.) In `main()` the function named `area()` is used twice: once to compute the area of a rectangle and once to compute the area of a square. In each case the function is different, but we can use the same name both times because we have defined `area()` to be a *virtual function* in the class `rectangle`. Then, which `area()` function is used depends on the class of the object when the function is called. This is an example of *polymorphism*, in which different meanings are assigned to the same identifier, depending on the context. Function overloading, as discussed in Chapter 14, is a primitive form of polymorphism in which the same name may be used for different functions. But with true polymorphism, the choice of the function is made at run time, not during compilation. This, in turn, is achieved by means of virtual functions.

In `main()` we first define `rec` and `sqr` to be objects of the classes `rectangle` and `square`, respectively, and then define a pointer to the class `rectangle`. We set

Calling the Functions

`pt` equal to the address of `rec` and then the statement

```
pt->read();
```

calls the member function `read()` in `rectangle`. This statement is equivalent to

```
(*pt).read();
```

and is an alternative way to call member functions when pointers are being used. Note that the correct `read()` function is called, rather than the one in `square`, because `pt` points to an object of the class `rectangle`. Similarly, the statement

```
pt->area()
```

EXAMPLE 19.2
Use of Virtual Functions

```cpp
#include <iostream.h>
class rectangle{ //define a class
   protected :
     float h, w; //height and width of rectangle
   public :
     virtual void read(){ //define a virtual function
       cout << "enter h and w\n";
       cin >> h >> w;
     }
     virtual void area(){ //define another virtual function
       cout << "area of rectangle = " << h * w << endl;
     }
};
class square : public rectangle{ //define a derived class
   private :
     float side; //side of square
   public :
     void read(){// define another read function
       cout << "side = ?\n";
       cin >> side;
     }
     void area(){ //compute area of square
       cout << "area of square = " << side * side << endl;
     }
};
int main(){
   rectangle rec; //define an object
   square sqr;     //define an object of the derived class
   rectangle *pt; //define a pointer to base class objects
   pt = &rec;      //pt points to rec
   pt -> read();   //call the read function in rectangle
   pt -> area();   //call the area function in rectangle
   pt = &sqr;      //pt now points to sqr
   pt -> read();   //call the read function in square
   pt -> area();   //call the area function in square
   return 0;
}
```

calls the `area` function from `rectangle`. Next, we set `pt` to the address of `sqr`. Now when we call the functions `read()` and `area()` in the next two statements, they are the functions from `square`. Thus it is the contents of `pt` that determines whether functions from `rectangle` or `square` are called. Without the use of virtual functions, any function calls on `pt->` would use functions from only the base class.

Functions are declared to be virtual only in the body of a class, and being virtual has no effect on a function unless a derived class uses the same function name. We note that constructors can't be virtual functions.

Abstract Classes

The classes in Example 19.2 are very similar. In fact, the area function of the square class is just a special case of the area function for the rectangle class. Suppose that instead of square, we wished another class, circle, that would compute the areas of circles. We could make circle a derived class of rectangle, or conversely, but this would violate the "is a" principle: A circle is not a rectangle

EXAMPLE 19.3
Use of an Abstract Base Class

```
#include <iostream.h>
class figure{//define base class
 public :
  virtual void read() = 0;//pure virtual function
  virtual void area() = 0;//pure virtual function
};
class rectangle : public figure{ //define a derived class
   private :
     float h, w; //height and width
   public :
     void read(){ //define read function
       cout << "enter h and w\n";
       cin >> h >> w;
     }
     void area(){ //define area function
       cout << "area of rectangle = " << h * w << endl;
     }
} ;
class circle : public figure{ //define another derived class
   private :
     float radius; //define read function
   public :
     void read(){ //define read function
       cout << "radius = ?\n";
       cin >> radius;
     }
     void area(){ //define area function
      cout << "area of circle = " << 3.14 * radius * radius << endl;
     }
};
int main(){
   rectangle rec;//define a rectangle object
   circle cir;   //define a circle object
   figure *pt = &rec;   //set a pointer to rectangle object
   pt -> read(); //call the read function in rectangle
   pt -> area(); //call the area function in rectangle
   pt = &cir;    //set a pointer to circle object
   pt -> read(); //call the read function in circle
   pt -> area(); //call the area function in circle
   return 0;
}
```

Figure 19.1 Circles and Rectangles Are Figures

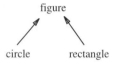

nor is a rectangle a circle. But both rectangles and circles are special cases of what we might call a figure, as depicted in Figure 19.1.

Figure 19.1 suggests that we define another class, `figure`, and use it as a base class for derived classes `rectangle` and `circle`. But what should the class `figure` consist of? The answer is shown in Example 19.3.

Pure Virtual Functions and Abstract Classes

In Example 19.3, we first define a base class `figure`. The functions `read()` and `area()` defined in this class are called *pure* virtual functions. They do not do anything within the base class and serve only to compel functions with the same names to be defined in derived classes. The `= 0` in the function declaration has nothing to do with assigning `0` to the function. This is simply the syntax for saying that the function is pure. The class `figure` in Example 19.3 is called an *abstract class* since it exists only as a base class for derived classes and no objects of the base class may be declared.

In Example 19.3, we define two derived classes: `rectangle`, which is the same as in Example 19.2 except that the member functions are not virtual, and `circle`, which has a member function to read a radius and a member function to compute the area of a circle. In `main()`, just as in Example 19.2, the proper `read()` and `main()` functions are called, depending on whether the pointer points to an object of `rectangle` or an object of `circle`. Note that, just as in Example 19.2, `pt` is a pointer to objects of the base class.

19.3 LINEAR EQUATIONS

We next give a much more significant and extensive example of the use of an abstract base class. As mentioned previously, many matrices that arise in applications are sparse; that is, most of the elements are zero. Tridiagonal matrices, discussed briefly in Chapter 12, are examples of such matrices. For tridiagonal matrices, a much more efficient form of Gaussian elimination may be used, as discussed in detail later in this section. Thus we would like to have two versions of Gaussian elimination, one for tridiagonal matrices and one for *dense* matrices, that is, matrices most of whose elements are nonzero. We can incorporate each of these different versions of Gaussian elimination into corresponding matrix classes. Then these two matrix classes can be derived from an abstract base class, as shown schematically in Figure 19.2.

The program to accomplish this is given in Example 19.4. The base class is called `MatrixBase` and contains three virtual functions. Then a derived class, `DenseMatrix`, is defined. This class is essentially the same as the matrix class of Example 18.11, except that we have added a print function and a function `solve()`

Figure 19.2 Different Derived Matrix Classes

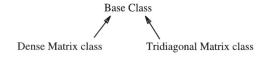

Base Class

Dense Matrix class   Tridiagonal Matrix class

that will solve the linear system $A\mathbf{x} = \mathbf{b}$ by Gaussian elimination. We leave as Exercise 19.5 to add a suitable copy constructor and functions for `operator=` and `operator[ ]` as in Example 18.11. The function `solve()` is the function `gauss()` of Exercise 16.4 except that the class `DenseMatrix` uses the class `vector` of Examples 18.3 and 18.4 [argumented by the `allot()` function of Example 18.10], and the arguments of `solve()` are references to vectors $\mathbf{x}$ and $\mathbf{b}$. (The size of the system is known from the protected variable `n` and does not need to be an explicit parameter.)

EXAMPLE 19.4
An Abstract Base Class for Solving Linear Systems

```
#include <iostream.h>
class MatrixBase {//An abstract base class
   public:
     virtual void read(int) = 0;
     virtual void print(int) = 0;
     virtual void solve(vector&,vector&) = 0;
};
class DenseMatrix : public MatrixBase{//A derived class
  //THIS CLASS USES THE VECTOR CLASS OF EXAMPLES 18.3 AND 18.4
  //AUGMENTED BY THE MEMBER FUNCTION ALLOT() OF EXAMPLE 18.10
   protected:
     int n;
     vector *A;//pointer to vectors
   public:
     DenseMatrix(){A = 0;} //default constructor
     DenseMatrix(int N);//constructor prototype
     ~DenseMatrix(){if(A != 0) delete [ ]A;} //destructor
     void read(); //prototype
     void print();//prototype
     void solve(vector&,vector&);//prototype for Gaussian elimination
};
DenseMatrix :: DenseMatrix(int N){//Constructor
   A= new vector[N];//allocate storage for pointers
   n = N;
   for(int i = 0;i < N;i++)//allocate storage for rows
     A[i].allot(N);
}
void DenseMatrix :: read(){
   cout << "input " << n << " squared values for matrix\n";
   for(int i = 0; i < n; i++)
     for(int j = 0; j < n; j++)
       cin >> A[i][j];
}
void DenseMatrix :: print(){
   cout << "this is the matrix\n";
   for(int i = 0; i < n; i++){
```

```
        for(int j = 0; j < n; j++) //display jth row
          cout << A[i][j] << " ";
        cout << endl; //new line after jth row
    }
}
//This is the Gaussian elimination function (see Example 13.1)
void DenseMatrix :: solve(vector &x, vector &b){
    for(int k =0; k < n; k++){
        float divisor = A[k][k];
        if(divisor == 0){ //check for zero divisor
          cout << "divisor is zero" << endl;
          return;
        }
        else //do triangular reduction
          for(int i = k + 1; i < n; i++){
            float multiplier = A[i][k]/divisor;
            for(int j = k + 1; j < n; j++)
              A[i][j] = A[i][j] -  multiplier * A[k][j];
            b[i] = b[i] - multiplier * b[k];
            }
    } //end of triangular reduction
    if(A[n-1][n-1] == 0){
        cout <<"Ann is zero" << endl;
        return;
    }
//now do back substitution
    for(int k = n - 1; k >= 0; k--){
      x[k] = b[k];
      for(int j = k + 1; j < n; j++)
        x[k] = x[k] - A[k][j] * x[j];
      x[k] = x[k]/A[k][k];
    }
}
//end of Gaussian elimination for dense matrices
//This class is for tridiagonal matrices. See text following program
class Tridiag : public MatrixBase{
    protected:
      int n;
      float *a; //pointer to main diagonal
      float *b; //pointer to super diagonal
      float *c; //pointer to subdiagonal
    public:
      Tridiag(int); //constructor prototype
      ~Tridiag(){delete [ ]a; delete [ ]b; delete [ ]c;} //destructor
      void read(int);
      void print(int);
      void solve(vector&,vector&);
};
Tridiag::Tridiag(int N){//constructor allocates storage for 3 diagonals
    a = new float[N];
    b = new float[N - 1];
    c = new float[N - 1];
    n = N;
}
void Tridiag :: read(){ //read 3 diagonals
```

```
      cout << "input " << n << " values for main diagonal\n";
      for(int i = 0; i < n; i++)
        cin >> a[i];
      cout << "input " << (n - 1) << " values for super diagonal\n";
      for(int i = 0; i < n -1; i++)
        cin >> b[i];
      cout << "input " << (n - 1) << " values for sub-diagonal\n";
      for(int i = 0; i < n -1; i++)
        cin >> c[i];
}
void Tridiag :: print(){ //print 3 diagonals
      cout << "this is the main diagonal of matrix\n";
      for(int i = 0; i < n;i++)
        cout << a[i] << " ";
      cout << "\nthis is the super-diagonal of matrix\n";
      for(int i = 0; i < n - 1;i++)
        cout << b[i] << " ";
      cout << "\nthis is the sub-diagonal of matrix\n";
      for(int i = 0; i < n - 1; i++)
        cout << c[i] << " "; cout << endl;
}
//Gaussian elimination for tridiagonal matrices. See text
void Tridiag :: solve(vector &x, vector &d){
      for(int k =0; k < n; k++){
        float divisor = a[k];
        if(divisor == 0){
          cout << "divisor is zero" << endl;
          return;
        }
        else{
          float multiplier = c[k]/divisor;
          a[k + 1] = a[k + 1] - multiplier * b[k];
          d[k + 1] = d[k + 1] - multiplier * d[k];
        }
      }
      if(a[n-1] == 0){
        cout << "an-1 is zero";
        return;
      }
//now do back substitution
      x[n - 1] = d[n - 1]/a[n - 1];
      for(int k = n - 2; k >= 0; k--)
        x[k] = d[k] - b[k] * x[k + 1];
}
int main(){
      cout << "N = ?\n";
      int N;
      cin >> N;
      DenseMatrix A(N);//declare a dense matrix
      MatrixBase *pt = &A;//set pointer to A
      cout << "A = ?\n";
      pt -> read();//input the matrix
      pt -> print();//display the matrix
      vector b(N),c(N),x(N);//declare three vectors
      cout << "b = ?\n";
```

```
b.read(); //input right hand side
b.print();//display right hand side
pt -> solve(x,b);//solve the system Ax = b
x.print();//display solution
Tridiag B(N);//declare a tridiagonal matrix
pt = &B;   //reset pointer to tridiagonal B
cout << "B = ?\n";
pt -> read();//input the tridiagonal matrix
pt -> print();//display the tridiagonal matrix
cout << "c = ?\n";
c.read(N);   //read right hand side
c.print(N); //display right hand side
pt -> solve(x,c);//solve the tridiagonal system Bx = C
x.print(N);//display solution
return 0;
}
```

A major point of Example 19.4 is that appropriate versions of the functions `read()`, `print()`, and `solve()` are used depending on the class of the object pointed to by `pt`. This is particularly powerful when `pt` is passed as a parameter to other functions that can then manipulate the matrices without knowing whether they are dense, tridiagonal, or something else.

Tridiagonal Matrices

After the class `DenseMatrix`, we define another class, `Tridiag`, also derived from the base class `MatrixBase`. This class solves linear systems $A\mathbf{x} = \mathbf{b}$ when A is a tridiagonal matrix as illustrated by:

$$A = \begin{bmatrix} a_1 & b_1 & & & \\ c_1 & a_2 & b_2 & & \\ & c_2 & \ddots & & \\ & & \ddots & & b_{n-1} \\ & & & c_{n-1} & a_n \end{bmatrix}$$

A tridiagonal matrix has nonzero elements only on its main diagonal and the two adjacent *super-* and *sub-*diagonals; all other elements in the matrix are zero.

Systems with a tridiagonal coefficient matrix can also be solved by Gaussian elimination, but the algorithm simplifies considerably because of all the zero elements. The simplified algorithm is shown by the pseudocode in Figure 19.3, for the system $A\mathbf{x} = \mathbf{d}$.

Advantages of Special Code for Tridiagonal Matrices

There are two major advantages in having a special program if Gaussian elimination is applied to tridiagonal systems. First, as mentioned at the end of Chapter 13, Gaussian elimination requires approximately $\frac{1}{3}n^3$ operations to solve an $n \times n$ system. But if the coefficient matrix is tridiagonal, only about $8n$ operations are required (Exercise 19.6). If n is large, this is a tremendous savings in operations; for example, if $n = 10^3$, $\frac{1}{3}n^3 = \frac{1}{3}10^9$, but $8n = 8000$. Second, only $3n - 2$ elements

Figure 19.3 Gaussian Elimination for a Tridiagonal System

for $k = 1, \ldots, n - 1$

    if $a_k = 0$ print error message and exit

    else

$$m_k = c_k / a_k$$
$$a_{k+1} = a_{k+1} - m_k b_k$$
$$d_{k+1} = d_{k+1} - m_k d_k$$

back substitution

$$x_n = d_n / a_n$$

for $k = n - 1, \ldots, 1$

$$x_k = d_k - b_k x_{k-1}$$

of the matrix need to be stored versus n^2 in general, again a tremendous savings in memory if n is large. In fact, it is customary to use three one-dimensional arrays to store the elements of the nonzero diagonals of the matrix, rather than use a two-dimensional array. This is the approach adopted in Example 19.4: The nonzero elements are stored in the three one-dimensional arrays a, b, and c. Thus the code for solve() in the class Tridiag follows the pseudocode of Figure 19.3 except that the indices run from 0 to $n - 1$. Note that we are assuming that the matrix is such that no interchanges are required in Gaussian elimination. A code with interchanges is somewhat more complicated, but roughly the same savings in computation and memory still apply.

Description of main()

In main() in Example 19.4, we first read in N, the size of the matrices and vectors, declare a matrix A(N) and set the pointer pt to the address of A. After reading the matrix, a vector b(N) is declared and read and then pt -> solve(x,b) calls the Gaussian elimination program to solve $A\mathbf{x} = \mathbf{b}$. Next, a tridiagonal matrix B(N) is declared and pt set to the address of B. Now, the statement pt -> read(N) asks us to input elements only for the three diagonals of the matrix. After a vector c(N) is read, the statement pt -> solve(x,c) calls the tridiagonal version of Gaussian elimination. The important point here is that in both cases, $A\mathbf{x} = \mathbf{b}$ and $B\mathbf{x} = \mathbf{c}$, a

Advantage of Abstract Base Class

function solve() is called, but because B(N) has been declared to be an object of the class Tridiag, the second call of solve() uses the member function solve() of the Tridiag class. Similarly for the functions read() and print(). This is the advantage of using an abstract base class.

Figure 19.2 and Example 19.4 have illustrated a fairly simple situation that could be expanded considerably. Figure 19.4 illustrates that in the MatrixBase class we could have a number of other virtual functions besides solve(). There might be one for addition of matrices, another for multiplication of matrices, and so on. Each of these other functions could have (and should have!) different implementations for dense and sparse matrices. And for sparse matrices we might wish to have separate classes for tridiagonal matrices and other more general sparse matrices.

Figure 19.4 Schematic of More General
Abstract Base Class

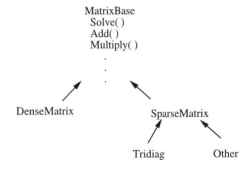

MAIN POINTS OF CHAPTER 19

- Classes may be derived from another class, called the base class. The derived class has access to the member variables and functions of the base class if these are public or protected. A derived class should be based on the `"is a"` principle: Any object of the derived class is also an object of the base class.

- Derived classes may also be base classes for still other derived classes.

- By the use of virtual functions, derived classes may use the same function names even though the function may be different in each class.

- An abstract base class exists only to allow derived classes, and no objects of the base class will be defined. Any functions in an abstract base class are pure virtual functions.

- By using derived classes and virtual functions, one can use base class pointers that point to derived class instances and call member functions without needing to know which function is actually invoked, or what class is doing the work. This facilitates the construction of programs and libraries that can operate on a wide variety of classes, without having to generate new code for each new type.

EXERCISES

19.1 Add implementations of the functions `print()`, `operator+`, and `operator*` to the `complex` class of the program of Example 19.1. Then run the program for various values of `alpha` and `beta` to test the error messages from the constructor, the operator= function and the read function. Next, delete one of the first two constructors in the definition of `ComplexPos` and note the syntax error. Do the same with the other constructor, after the first one has been replaced.

19.2 Add a member function to do division of complex numbers (Exercise 17.2) to the class `ComplexPos` of Example 19.1. Use the result of a division in an assignment statement to verify that the operator= function detects numbers without positive real part in this case also.

19.3 The area of an ellipse whose major axis has length a and whose minor axis has length b is $ab\pi/4$. Modify Example 19.2 so that the class `rectangle` is replaced by a new class `ellipse` and the derived class `square` by a new derived class `circle`.

19.4 Define an abstract base class as in Example 19.3 and then define the classes `rectangle` of Example 19.2 and `ellipse` of Exercise 19.3 as classes derived from the abstract base class. Next, define classes `square` and `circle` as derived classes of `rectangle` and `ellipse`, respectively. You should now have four different `read()` and `area()` functions. Write a function `main()` that calls these different functions.

19.5 Write a copy constructor, an `operator=` function and an `operator[ ]` function for the `DenseMatrix` class of Example 19.4. Do the same for the `Tridiag` class.

19.6 Count the arithmetic operations required to solve an $n \times n$ tridiagonal linear system by Gaussian elimination.

19.7 The *transpose* of a matrix $A = (a_{ij})$ is the matrix $A^T = (a_{ji})$; that is, the elements of A^T are those of A "flipped" across the main diagonal. A matrix A is *orthogonal* if $A^T A = I$, the identity matrix, which has ones on the main diagonal and zeros elsewhere. If A is an orthogonal matrix, the solution of the linear system $Ax = \mathbf{b}$ is given by $\mathbf{x} = A^T \mathbf{b}$. Add another derived class to solve systems with an orthogonal coefficient matrix to the program of Example 19.4. Verify that the matrix

$$\begin{bmatrix} \cos\theta & \sin\theta \\ -\sin\theta & \cos\theta \end{bmatrix}$$

is orthogonal for any angle θ and use such matrices to test your program.

FURTHER READING

There is a large, and growing, number of introductory books on C++. Here are two:

J. Cohoon and J. Davidson, *C++ Program Design: An Introduction to Programming and Object-Oriented Design*, Irwin, 1996.

W. Savitch, *Problem Solving with C++*, Addison-Wesley, 1996.

A more advanced book by the originator of the C++ language is

B. Stroustrup, *The C++ Programming Language*, 2nd Edition, Addison-Wesley, 1995.

None of the above books has any particular emphasis on scientific computing and numerical methods. The following more advanced books do:

J. Barton and J. Nackman, *Scientific and Engineering C++*, Addison-Wesley, 1994.

D. Capper, *Introducing C++ for Scientists, Engineers and Mathematicians*, Springer-Verlag, 1994.

Our introduction to numerical methods for scientific computing has necessarily been at an elementary level, and some of the methods discussed are not very good for most applications. This is especially true of the methods for numerical integration and for differential equations. More advanced discussions of these topics are given in a number of books devoted to numerical methods; two references are:

R. Burden and J. Faires, *Numerical Analysis*, 4th ed, PWS-Kent Publishing Company, 1989.

G. Golub and J. Ortega, *Scientific Computing and Differential Equations*, Academic Press, 1992.

Although our treatment of Gaussian elimination is reasonably complete, at least at this level, the reader is advised to use the programs in LAPACK if available. LAPACK replaces LINPACK, which has been the standard package of linear equation solvers for over a decade. A reference is:

E. Anderson, Z. Bai, C. Bischof, J. Demmel, J. Dongarra, J. DuCroz, A. Greenbaum, S. Hammarling, A. McKenney, S. Ostrouchov, and D. Sorensen, *LAPACK User's Guide*, SIAM, 1992.

Appendix 1

ASCII Character Codes

Character	Binary (Decimal)	Character	Binary (Decimal)	Character	Binary (Decimal)
A	1000001 (65)	a	1100001 (97)	0	0110000 (48)
B	1000010 (66)	b	1100010 (98)	1	0110001 (49)
C	1000011 (67)	c	1100011 (99)	2	0110010 (50)
D	1000100 (68)	d	1100100 (100)	3	0110011 (51)
E	1000101 (69)	e	1100101 (101)	4	0110100 (52)
F	1000110 (70)	f	1100110 (102)	5	0110101 (53)
G	1000111 (71)	g	1100111 (103)	6	0110110 (54)
H	1001000 (72)	h	1101000 (104)	7	0110111 (55)
I	1001001 (73)	i	1101001 (105)	8	0111000 (56)
J	1001010 (74)	j	1101010 (106)	9	0111001 (57)
K	1001011 (75)	k	1101011 (107)		0100000 (32)
L	1001100 (76)	l	1101100 (108)	!	0100001 (33)
M	1001101 (77)	m	1101101 (109)	"	0100010 (34)
N	1001110 (78)	n	1101110 (110)	#	0100011 (35)
O	1001111 (79)	o	1101111 (111)	$	0100100 (36)
P	1010000 (80)	p	1110000 (112)	%	0100101 (37)
Q	1010001 (81)	q	1110001 (113)	&	0100110 (38)
R	1010010 (82)	r	1110010 (114)	'	0100111 (39)
S	1010011 (83)	s	1110011 (115)	(	0101000 (40)
T	1010100 (84)	t	1110100 (116)	)	0101001 (41)
U	1010101 (85)	u	1110101 (117)	*	0101010 (42)
V	1010110 (86)	v	1110110 (118)	+	0101011 (43)
W	1010111 (87)	w	1110111 (119)	,	0101100 (44)
X	1011000 (88)	x	1111000 (120)	-	0101101 (45)
Y	1011001 (89)	y	1111001 (121)	.	0101110 (46)
Z	1011010 (90)	z	1111010 (122)	/	0101111 (47)
[	1011011 (91)	{	1111011 (123)	:	0111010 (58)
\	1011100 (92)	|	1111100 (124)	;	0111011 (59)
]	1011101 (93)	}	1111101 (125)	<	0111100 (60)
∧	1011110 (94)			=	0111101 (61)
_	1011111 (95)			>	0111110 (62)
`	1100000 (96)			?	0111111 (63)
				@	0100000 (64)

Appendix 2

Library Functions

We summarize in this appendix some of the most useful functions in the `math.h` library. Many of these have been discussed in the text. All these functions have return type `double` and accept `double` parameters. However, they can be used with `float` parameters, which are then automatically cast as double.

fabs(x)	Returns $|x|$
sqrt(x)	Returns $\sqrt{x}$, x must be nonnegative
pow(x, a)	Returns x^a, x must be nonnegative, if not integer
exp(x)	Returns e^x
log(x)	Returns $\ln(x)$
log10(x)	Return $\log_{10}(x)$
sin(x)	Returns $\sin(x)$
cos(x)	Returns $\cos(x)$
tan(x)	Returns $\tan(x)$
asin(x)	Returns $\arcsin(x)$
acos(x)	Returns $\arccos x$
atan(x)	Returns $\arctan(x)$
sinh(x)	Returns hyperbolic sin of x
cosh(x)	Returns hyperbolic cosine of x
tanh(x)	Returns hyperbolic tangent of x
ceil(x)	Returns smallest integer greater than or equal to x
floor(x)	Returns largest integer less than or equal to x

INDEX